P9-ARF-728

THINKING VISUALLY

THINKING VISUALLY

A STRATEGY MANUAL FOR PROBLEM SOLVING

Robert H. McKim

LIFETIME LEARNING PUBLICATIONS a division of Wadsworth, Inc. Belmont, California

Designer: Joe di Chiarro

Jacket Design: Sam Smidt

© 1980 by Wadsworth, Inc. All rights reserved. No part of
this book may be reproduced, stored in a retrieval system,
or transcribed, in any form or by any means, electronic,
mechanical, photocopying, recording, or otherwise,
without the prior written permission of the publisher,
Lifetime Learning Publications, Belmont, California 94002,
a division of Wadsworth, Inc.

Printed in the United States of America

3 4 5 6 7 8 9 10—84 83

Library of Congress Cataloging in Publication Data
McKim, Robert H
  Thinking visually.

  Bibliography: p.
  Includes index.
  1. Problem solving.   2.   Thought and thinking.
3. Visualization.   I.   Title.
BF441.M19   153.3′2   80–16526
ISBN 0–534–97984–X

To Virginia

PREFACE

Although *Thinking Visually* is written for anyone who thinks visually (and that is virtually everyone), it is most clearly intended for people in the visual professions—designers, architects, engineers, and artists—people who are confronted every day with visual-spatial problems. An unresolved floorplan; a needed ingenious mechanism; a flawed artistic composition: if you are already a seasoned and successful solver of such visual braintwisters, this book will enhance your professional creativity by helping you to identify and strengthen the problem-solving strategies that you currently use while also showing you the power of new ones. If you are a beginner in a visual profession, *Thinking Visually* will introduce you to an exciting and effectual way to accelerate your professional growth.

The usefulness of this book to professional problem-solvers in nonvisual fields is perhaps more important, though less obvious. People who find themselves bogged down in the verbal patterns of their profession will find visual thinking to be an invigorating and liberating antidote. Thus while *Thinking Visually* has patent utility for designers and engineers, it may have even more potent utility for problem-solvers in fields such as law, psychology, business, or education, where thinking is often unduly constrained by the limits of language.

Thinking Visually is a *do-it* book, a provocation to explore, to experience, and to use your mental powers to the full in the real world of problem-solving. It is also a *how-to* book, in the sense that it points the way for you to more fully realize your untapped thinking abilities. I have endeavored to present this challenge to think visually in simple, nontechnical language supported with abundant illustrations and numerous opportunities for you to experience ideas about thinking *in your own thinking*. And finally, I have turned *Thinking Visually* into a strategy manual that you can use as a handy, everyday guide to problem-solving.

I hope that you will find this book thought-provoking and useful. I will also welcome hearing from you, especially about your personal experiences in the fascinating realm of visual thinking.

Robert H. McKim

CONTENTS

TWO WAYS TO APPROACH THIS BOOK

This book contains two ways to improve your ability to think visually:

1. An *experiential skills approach* designed to help you acquire skills in visual thinking by performing numerous short exercises.

2. A *strategy approach* designed to help you use your newly experienced visual-thinking skills to solve actual problems.

Although you may want to start immediately to use the book to solve problems, I recommend that you begin with the experiential skills approach: you will be far more effective in using the strategy approach to problem-solving when you have a complete sense of the many and varied strategies available to you.

The Experiential Skills Approach

At the heart of the experiential skills approach to the book are the numbered sequences that appear in each chapter: exercises, puzzles, and thinking challenges designed to illustrate and elicit specific kinds of visual thinking. The first purpose of this approach is to enable you to experience ideas about thinking in your own thinking. Its second purpose is to provide a means for you to exercise your thinking skills. No skill, whether it be skill in basketball, in playing the cello, or in thinking, can be acquired by passive reading; skills can be acquired only by active and informed experience.

The value of the experiential skills approach will become apparent to you as you proceed into Section I, Background. The first chapter leads you to experience your own thinking, and on the basis of this experience, to obtain a direct and personal sense of the value of thinking visually. Subsequent chapters in this section expand on three basic kinds of imagery used in visual thinking, ways that this imagery can be manipulated and transformed, and the role that image thinking plays in healthy neurological functioning.

Section II, Preparations, introduces you to materials especially suited to visual-thinking activity and encourages you to organize an environment that is conducive to productive thinking. In addition to these physical preparations, Section II guides you to a mental preparation that facilitates all thinking activity: relaxed attention.

Section III, Seeing, is devoted to vitalizing vision and to integrating the visual sense with thinking. You begin by experiencing "externalized thinking," thinking in the direct context of seeing. Then the importance of recentering one's viewpoint is stressed; creative vision accompanies the ability to see from new vantage points. Experiences in drawing afford another way to invigorate vision. By drawing, you will experience the pattern-seeking as well as the analytical nature of seeing, and you will explore perceptual cues that make your visual world three-dimensional.

In Section IV, Imagining, your attention is turned inward. Here you experience various kinds of inner imagery, from autonomous dream and hypnogogic imagery, to directed fantasy, to logical, abstract, and structural imagery. The thrust of these experiences is to open up inner imaginative resources essential to creative visual thinking.

Section V, Idea-Sketching, directs your attention outward again. To "express" means to "press out"; in this section, you will learn to express your visual ideas on paper, using the graphic language that is most appropriate to the idea's level of abstractness. Other strategies for thinking with your pencil, such as visual brainstorming and development-by-overlay, are also presented.

The Strategy Approach

As you experience the numbered sequences in each chapter, you will realize that each is also a strategy that can be used to solve problems. In Chapter 22, The Strategy Approach, experiential and skills-oriented exercises are reformulated as problem-solving strategies. Grouped into visual, easily accessible form, your initial skill experiences are thus transformed into a set of powerful tools for solving problems.

The strategy approach presented in this book is emphatically not intended to be used as a step-by-step problem-solving method. Effective problem-solving does not follow a cookbook pattern, such as

1. Recenter your viewpoint.

2. Add a dab of inner imagery.

3. Now make six small idea-sketches.

4. Cook in the brain overnight.

Quite the opposite, visual thinking that is equipped with a wide range of responses moves *flexibly* to solve problems. The strategy approach is intended to help you choose another approach when you are stuck on a problem. When seeking another direction to take, simply refer to the Strategy Flow Chart and Strategy

Index (following Chapter 22) for an overview of the alternative thinking strategies available to you.

The Pragmatic Test

Because the subjective nature of thinking tends to elude scientific observation, students of thinking are faced with a subject about which science has learned little. Consequently, this book takes a path to understanding and skill that is essentially personal and pragmatic. It leads you, the reader, to be a rigorous investigator of your own thinking, to test ideas about thinking in the laboratory of your mind, and to repeatedly ask the question, "Does this particular assumption about thinking work for me?"

This pragmatic approach to self-teaching requires that you maintain a balance between healthy skepticism and openness to new experience. And, as with skill acquisition generally, it may also require that you exercise considerable patience. A natural tendency to resist changes in your current thinking pattern can be *prematurely misinterpreted* as failure of a given skill to pass the test of "Does it work for me?" Acquiring new skills in any field requires sustained and repeated effort.

In brief, though I believe that skill in visual thinking is available and vitally important to virtually everyone, I cannot validate this claim for you. Only you can. Anything so fundamental to your being, so powerful in its capacity to change your world of ideas and things, cannot come off the printed page and into your productive reality without your active and sustained effort. Visual thinking develops from your unique experience. No two people ever see, imagine, or draw the same; no two people ever think the same thoughts. I have tried only to provide an initial stimulus. The rest is up to you.

THINKING VISUALLY

BACKGROUND

1

The following four chapters provide background about the nature of visual thinking. The first chapter encourages you to explore several basic attributes of your own thinking, particularly that of flexibility, and to experience the importance of thinking visually to fully effective problem-solving. The second chapter describes three kinds of visual imagery that are the primary vehicles of visual thinking. The third chapter shows how visual thinking involves many kinds of *active* mental operations. The fourth chapter discusses ways in which visual thinking provides an important and creative complement to modes of thinking such as verbal thinking.

1. Thinking Flexibly
2. Thinking by Visual Images
3. Images in Action
4. Ambidextrous Thinking

THINKING FLEXIBLY

Many words link vision with thinking. *Insight, foresight, hindsight,* and *oversight. Visionary* and *seer.* The word "idea" derives from the Greek "idein": to see. A sound thinker is *sens*-ible, or possesses common *sense*; a creative thinker is *imaginative* or *farsighted,* a productive *dreamer.* Common phrases also connect thinking to the visual sense: *See* what I mean? *Look* at the idea from another *viewpoint.* Before *focusing* in, examine the *big picture.* Indeed, before entering further into a discussion about visual thinking, let us step back, take a larger *perspective,* and consider thinking generally.

What Is Thinking?

The word "thinking" is much used but little understood. When do you think? Do you think rarely, or most of the time? Do you think when you are asleep? Does thinking occur only in your brain, or does it extend out into your nervous system, sense organs, even into your muscles? Is the raw material of thinking a kind of inner speech, sensory imagery, or is it an imageless process that occurs below the threshold of your consciousness?

Look at the photograph of Rodin's statue "The Thinker" in Figure 1-1. For all we know, Rodin's model is daydreaming, suffering a hangover, or taking a nap; he may even be in a psychotic state of catatonic withdrawal. The most obvious and at the same time most important observation about thinking is that thinking is extremely difficult to observe. Clearly, it cannot be observed in the same way that a scientist observes external physical phenomena. Only thinkers themselves can know what they are thinking, and not even they can fully know.

Despite its elusiveness, thinking is extremely pervasive. Measurements of the brain's electrical activity, for example, show almost constant mental activity, even when the subject is asleep. You can experience this incessant mentation for yourself by attempting to stop it. Close your eyes and try to silence all inner speech and to blank out all inner imagery. For five minutes, sustain a quiet mind that does not think. You will likely find this instruction much easier read than done.

Thinking not only occupies most of

Figure 1-1.

your waking and sleeping time; it also utilizes most of your being. Mental functions cannot be readily separated from bodily ones. Physiologists have shown that muscle tone plays a part in mental functioning (see Chapter 6), neurologists will tell you that the entire nervous system (not just the brain) is involved in thinking, and you know by experience that the vitality of your thinking is intimately related to the state of your physical health. Also, thinking is never totally divorced from feeling. Your thoughts are always colored and directed by your emotions and motivations.

So far, I have suggested that thinking, while not easily observed, is a constant fact of life. Irrational or sane, habit-ridden or brilliantly incisive, logical or illogical, awake or dreaming, we think with our entire being almost all of the time. By this broad definition, most thinking is *not* productive. We need not assume that mental activity which merits being called thinking is necessarily good thinking. Indeed, most thinking that is eventually productive is preceded by frustrating, cyclic, abortive, ill-informed, illogical, habit-plagued thinking that produces (at the time) very little of value.

Self-Observation

A fascinating way to learn about thinking is to watch your own thinking in action. In the following experience, you will find that observing your own thinking is a demanding task: while thinking, you must also somehow step to one side and *watch your thinker thinking*.

1•1 MONK-ON-THE-MOUNTAIN

As you think to solve the following puzzle, observe your thoughts to the best of your ability:

"One morning, exactly at sunrise, a Buddhist monk began to climb a tall mountain. The narrow path, no more than a foot or two wide, spiraled around the mountain to a glittering temple at the summit.

"The monk ascended the path at varying rates of speed, stopping many times along the

way to rest and to eat the dried fruit he carried with him. He reached the temple shortly before sunset. After several days of fasting and meditation, he began his journey back along the same path, starting at sunrise and again walking at variable speeds with many pauses along the way. His average speed descending was, of course, greater than his average climbing speed.

"Prove that there is a single spot along the path the monk will occupy on both trips at precisely the same time of day."[1]

Before coming to the answer, let us discuss what you observed of your thinking; from your self-observations, we should be able to draw three basic conclusions about the general nature of the thought process.

First, did you notice how your thinking was represented to your consciousness? You may have been aware, for example, that you talked to yourself subvocally, with a kind of inner speech. Or perhaps you mentally pictured an image of the monk climbing to the top of the mountain. Or, possibly, you drew a diagram to clarify your thinking. Whatever you observed of your mental processes, we will call the *vehicle* of your thinking. A vehicle of thought can be a feeling, a bodily gesture, a mathematical notation, a sensory image—however your thinking is represented to your consciousness. *To repeat, a vehicle is a representation of thought.* It is, therefore, a result of thought, and not thought itself. To come closer to the very nature of thinking, we must return to self-observation for clues.

As you reflect on your puzzle-solving, you probably will be aware that you removed some of the extraneous details (the dried fruit, for example, or the "glittering" temple) and simplified essential elements (possibly by representing the mountain path as a line in space). In other words, you acted on your thoughts, you reformulated what you read, by abstracting out the essential elements of the puzzle. This observation leads us to a second conclusion about thinking generally: we think by performing a number of active mental *operations*.

Another active thinking operation frequently performed with this puzzle is a "rotation" in which the mountain is viewed from above and the path is seen as a spiral. The operation that solves the puzzle is "superimposition": an image of the monk-of-the-first-day is superimposed on an image of the monk-of-the-second-day, revealing that the ascending monk must confront the descending monk on a single spot on the path, obviously at the same time of day. Figure 1-2 illustrates the solution.

So far, you have seen that thinking is represented to your consciousness by *vehicles* of thought, and that thinking itself occurs by varied and active mental *operations*. A third conclusion about the nature of thinking follows when you reflect on how much of your thinking eluded your attempts at self-observation. For example, did you consciously choose to abstract the puzzle by removing extraneous details, or did you become aware of performing this operation only after the fact? Did you consciously choose to superimpose the two images, or did you infer from the resulting image that you had performed this mental operation subconsciously? Most people

infer operations by observing the resulting vehicle. Operations themselves are usually chosen and performed below the level of conscious awareness. Indeed, most of your thinking occurs at a subconscious level, a fact of mental life that makes thinking extremely difficult to study. So, to *vehicles* and *operations* of thinking, we add *levels:* some of our thinking occurs above, and most of it below, the level of our conscious awareness.

In addition to *vehicles, operations,* and *levels* of thinking (which are characteristics of all thinking, whether it is productive or not), we can identify three conditions that clearly promote effective thinking. The first is *challenge:* we think at our best when posed with a situation that we deeply desire to change. The second is *information:* since thinking is essentially information-processing, we cannot expect productive thinking when information is incorrect, inadequate, or tucked away in an unavailable crevice of memory. Each reader must seek these first two conditions without much aid from this book: challenge is a personal equation, and information requirements vary with each problem.

A major purpose of this book is to encour-

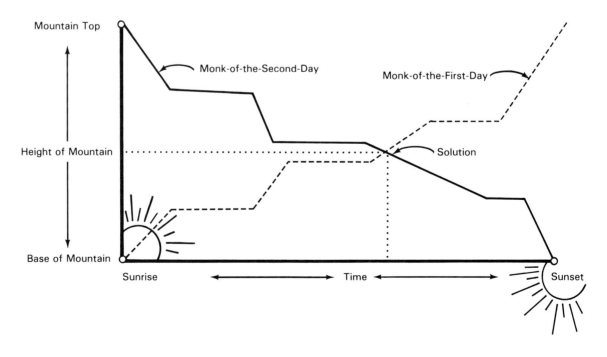

Figure 1-2.

age a third universal condition that fosters productive thinking: flexibility. Productive thinkers can be flexible in their thinking in three ways:

1. They can have easy access to subconscious as well as to conscious *levels* of thinking.

2. They can be proficient at a variety of mental *operations* and able to move freely from one operation to another.

3. They can utilize several *vehicles* of thought, and readily transfer their thinking from one vehicle to another.

What is meant by flexibility in levels, operations, and vehicles of thinking?

Flexibility in Levels

Our flow of thoughts has often been compared to a stream: "We never have single thoughts but always many, an entire polyphony. . . . I picture thinking as a stream of which only the surface is visible, orchestral music of which only the melody is audible."[2] Another common image for thinking is that of an iceberg: the small, visible portion of the iceberg is comparable to the mental processes of which we are consciously aware, and the much larger, submerged portion, to the unconscious mental activity that by definition goes unobserved.

You are demonstrating flexibility in levels of thinking when you decide to stop thinking consciously about a problem, to relax, to take a walk, to sleep on it—in short, to allow your thinking to proceed unconsciously. Flexible thinkers are also alert to recognize ideas that emerge from unconscious levels. In Chapter 16, for example, I discuss a number of creative insights obtained from dreams.

Education rarely encourages flexibility in levels of thinking. In school conscious modes of thinking are usually stressed, and subconscious modes are rarely even mentioned. One purpose of this book is to point to this educational oversight and to suggest ways in which you can become aware of and tap into thinking that occurs below the threshold of your normal waking consciousness.

Flexibility in Operations

Earlier in this chapter, abstraction, rotation, and superimposition were given as examples of the active operations you perform when you think. Another example of a thinking operation is *analysis:* by analysis, you actively dissect the object of your thinking into parts (for example, the a, b, c of an outline). A distinctly different operation is *synthesis:* by synthesis, you actively combine two or more unlike ideas into a new entity (as in an invention). Still another operation is *induction:* by induction, you move from particular observations to a generalized concept. *Deductive* thinking operates in the opposite direction, from the general to the particular.

Most thinkers are disposed to use a limited set of favorite thinking operations. The logical thinker likes to operate by rules of logic, step by step, in a single direction. The intuitive thinker, by contrast, prefers to take "mental leaps," often in surprising directions.

As a skillful carpenter masters many tools and, matching tool to task, expertly moves from one tool to another, so the productive thinker has flexible access to a wide variety of mental operations. An important purpose of this book is to encourage you to enlarge your working repertoire of thinking operations and to learn the value of moving from one operation to another.

Flexibility in Vehicles

While thinking operations are the form of action that thinking takes, thinking vehicles are, as you experienced a moment ago, the means by which this action is represented to consciousness. A well-known thinking vehicle is language: as one writer bluntly states, "Thought is born through words."[3] Other vehicles of thinking are nonverbal languages (such as mathematics), sensory imagery, and feelings.

Being able to move flexibly from one thinking vehicle to another is frequently essential when we move from one kind of problem to another. An architect, for example, will quickly find thinking blocked if forced to design a building exclusively with words. To have an analogous experience of frustration, try to describe even the simplest object over the telephone.

The real power of flexibility in thinking vehicles, however, lies not in the vehicle but in the operations made possible by the vehicle. The Monk-on-the-Mountain puzzle dramatizes this important point. A thinker who chooses to solve this puzzle by the vehicle of language is doomed to fail, because the sequential operations built into the syntax of language, while extremely valuable for certain kinds of thinking, do not include the operation of superimposition. The forging of two images into one is an operation available only to the thinker using the vehicle of visual imagery.

When considering this argument for flexibility in choice of thinking vehicles, remember that vehicles are representations of thought, not thought itself. Consequently, turning your thinking to a particular vehicle opens the door to mental operations associated with that vehicle, but does not necessarily cause the operations to enter into your thinking. On the other hand, rigidly adhering to one vehicle does tend to close doors. This book strongly promotes an "open-door policy." As we will discuss in the remainder of this section, the visual vehicle, with its link to holistic, spatial, metaphoric, transformational operations, provides a vital and creative complement to the reasoning, linear operations built into the vehicle of language.

References

1. Duncker, C. Quoted by Koestler, A., in *The Act of Creation.* Macmillan.
2. Stekel, W. Quoted by McKellar, P., in *Imagination and Thinking.* Basic Books.
3. Vygotsky, L. *Thought and Language.* The M.I.T. Press.

Recommended Reading

For a thorough and readable treatment of thinking in general, I suggest *A Study of Thinking* by Jerome S. Bruner, Jacqueline J. Goodnow, and George A. Austin (Wiley). For a discussion of subconscious levels of thinking, I recommend Erich Fromm's *The Forgotten Language* (Grove Press), P. W. Martin's *Experiment in Depth* (Pantheon), Carl Gustav Jung's *Man and His Symbols* (Dell), and Lawrence Kubie's *Neurotic Distortion of the Creative Process* (Farrar, Straus & Giroux–Noonday Press). Thinking operations are thoroughly treated by J. P. Guilford in *The Nature of Human Intelligence* (McGraw-Hill). Vehicles of thinking are discussed in the books by McKellar and Vygotsky listed above, and from the standpoint of general semantics, by S. I. Hayakawa in *Language in Thought and Action* (Harcourt Brace Jovanovich).

The topic of thinking can be approached by many other avenues. For a treatment of how thinking ability is gradually developed, I recommend Jean Piaget's and Bärbel Inhelder's *The Child's Conception of Space* (Norton). Thinking is fascinatingly treated as a neurophysiological function by W. Grey Walter in *The Living Brain* (Norton). George Miller, Eugene Galanter, and Karl Pribram, in *Plans and the Structure of Behavior* (Holt, Rinehart & Winston), combine insights into thinking from such fields as cybernetics, neurology, computer science, and psychology. All of these books are understandable by the lay reader. For an excellent and much larger bibliography, look in the back of Bruner, Goodnow, and Austin's *A Study of Thinking* (Wiley).

THINKING BY VISUAL IMAGES

2

Visual Thinking Is Pervasive

Visual thinking pervades all human activity, from the abstract and theoretical to the down-to-earth and everyday. An astronomer ponders a mysterious cosmic event; a football coach considers a new strategy; a motorist maneuvers his car along an unfamiliar freeway: all are thinking visually. You are in the midst of a dream; you are planning what to wear today; you are making order out of the disarray on your desk: *you* are thinking visually.

Surgeons think visually to perform an operation; chemists to construct molecular models; mathematicians to consider abstract space-time relationships; engineers to design circuits, structures, and mechanisms; administrators to organize and schedule work; architects to coordinate function with beauty; carpenters and mechanics to translate plans into things.

Visual thinking, then, is not the exclusive reserve of artists. As Rudolf Arnheim observes, "Visual thinking is constantly used by everybody. It directs figures on a chessboard and designs global politics on the geographical map. Two dexterous moving-men steering a piano along a winding staircase think visually in an intricate sequence of lifting, shifting, and turning. . . . An inventive housewife transforms an uninviting living room into a room for living by judiciously placing lamps and rearranging couches and chairs."[1]

See/Imagine/Draw

Visual thinking is carried on by three kinds of visual imagery:

1. The kind that we *see:* "People see images, not things."[2]

2. The kind that we *imagine* in our mind's eye, as when we dream.

3. The kind that we *draw,* doodle, sketch, or paint.

Although visual thinking can occur primarily in the context of seeing, or only in imagination, or largely with pencil and paper, expert visual thinkers flexibly utilize all three kinds of imagery. They find that seeing, imagining, and drawing are interactive.

Interactive Imagery

The interactive nature of seeing, imagining, and drawing is shown diagrammatically in Figure 2-1. The overlapping circles can be taken to represent a wide variety of interactions. Where seeing and drawing overlap, seeing facilitates drawing, while drawing invigorates seeing. Where drawing and imagining overlap, drawing stimulates and expresses imagining, while imagining provides impetus and material for drawing. Where imagining and seeing overlap, imagination directs and filters seeing, while seeing, in turn, provides raw material for imagining. The three overlapping circles symbolize the idea that *visual thinking is experienced to the fullest when seeing, imagining, and drawing merge into active interplay.*

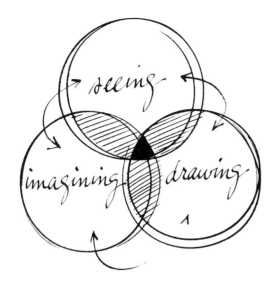

Figure 2-1.

The visual thinker utilizes seeing, imagining, and drawing in a fluid and dynamic way, moving from one kind of imagery to another. For example, you might *see* a problem from several angles and perhaps even choose to solve it in the direct context of seeing. Now prepared with a visual understanding of the problem, you *imagine* alternative solutions. Rather than trust to memory, you *draw* a few quick sketches, which you can later evaluate and compare. Cycling between perceptual, inner, and graphic images, you continue until the problem is solved.

Experience this interplay between perceptual, inner, and graphic images for yourself, as you solve a challenging and somewhat difficult classic puzzle (2-1).

Use seeing, imagining, and drawing to solve this problem, as follows:

1. Simulate the pierced block with a cardboard cut-out. With scissors and cardboard, seek to see a solution by actual "cut-and-try" methods.

2. Close your eyes and seek a solution in your imagination.

3. Make sketches; seek a graphic solution.

4. Consciously alternate between steps 1, 2, and 3.

(An answer to this puzzle is illustrated at the end of this chapter.)

Visual thinking is obviously central to the practice of architecture, design, and the visual arts. Less obvious is the importance of visual thinking to other disciplines, such as science and technology. In the next few pages I will present a few brief accounts of seeing, imagining, and drawing in the thinking of scientists and technologists. Interspersed with these, I have placed related problems that will help you to relate these accounts of others to your own experience.

Seeing and Thinking

Discoveries in the direct context of seeing are common in the history of science. For example, Sir Alexander Fleming noticed, while working with some plate cultures of staphylococci, that one colony of bacteria had apparently become contaminated and died. Although most bacteriologists would

---

2•1 PIERCED BLOCK

Figure 2-2 shows a solid block that has been pierced with circular, triangular, and square holes. The circle's diameter, the triangle's altitude and base, and the square's sides all have the same dimension. The walls of the three holes are perpendicular to the flat front face of the block. Visualize a single, solid object that will pass *all the way through* each hole and, en route, entirely block the passage of light.

Figure 2-2.

have overlooked it, knowing that some bacteria can interfere with the growth of others, Fleming saw it in a way that enabled him to transform a routine laboratory accident into a profound scientific event, the discovery of penicillin.

Why did Fleming discover penicillin when another scientist saw the same thing and thought it a nuisance? Because *habits of seeing and thinking are intimately related.* Fleming, like most creative observers, possessed a habit of mind that permitted him to see things afresh, from new angles. Also, he was not burdened by that "inveterate tradition according to which thinking takes place remote from perceptual experience."[1] He did not look and *then* sit down to think; he used his active eyes and mind *together.*

Experience using your eyes and mind together in puzzle 2-2.

2•2 CARDS AND DISCARDS[3]

"In [the] row of five cards shown [in Figure 2-3], there is only one card correctly printed, there being some mistake in each of the other four. How quickly can you find the mistakes?"

Watson and his colleagues visualized this complex structure by interacting directly with a large three-dimensional model. He writes:

> Only a little encouragement was needed to get the final soldering accomplished in the next couple of hours. The brightly shining metal plates were then immediately used to make a model in which for the first time all the DNA components were present. In about an hour I had arranged the atoms in positions which satisfied both the x-ray data and the laws of stereochemistry. The resulting helix was right-handed with the two chains running in opposite directions. . . . Another fifteen minutes' fiddling by Francis [Crick] failed to find anything wrong, though for brief intervals my stomach felt uneasy when I saw him frowning.[4]

Although a complex structure such as the DNA molecule is difficult to visualize in imagination or on paper, one of Watson's colleagues scorned the model shown in Figure 2-4. However, as Watson observed, his Nobel Prize-winning success by this method of visual thinking convinced the doubter "that our past hooting about model-building represented a serious approach to science, not the easy re-

Figure 2-3.
From *The Book of Modern Puzzles* by Gerald L. Kaufman, Dover Publications, Inc., New York, 1954. Reprinted through permission of the publisher.

Another form of thinking in the context of seeing is described by Nobel Laureate James D. Watson in *The Double Helix,* a fascinating account of the discovery of the structure of the DNA molecule.

sort of slackers who wanted to avoid the hard work necessitated by an honest scientific career."

Watson's account in *The Double Helix* also gives the reader excellent insight into the competitive excitement of science. Ideally, the next problem (2-3), an experience in thinking in the direct context of seeing, should be approached in the spirit of a competition.

2•3 SPAGHETTI CANTILEVER

With 18 sticks of spaghetti and 24 inches of Scotch tape, construct the longest cantilever structure that you can. Here are three additional constraints:

1. Tape-fasten the base of the structure within a 6-inch-square horizontal area.

2. Don't make drawings. Think directly with the materials.

3. Design and build the structure in 30 minutes.

(Measure length of cantilever from the point on the base nearest to the overhanging end of the cantilever.)

The relation between seeing and thinking will be treated further in Section III, Seeing.

Imagining and Thinking

Inner imagery of the mind's eye has played a central role in the thought processes of many creative individuals. In rare thinkers, this inner imagery is extremely clear. For example, Nikola Tesla, the technological genius whose list of inventions includes the fluorescent light and the A-C generator, "could project before his eyes a picture, complete in every detail, of every part of the machine. These pictures were more vivid than any blueprint."[5] Tesla's inner imagery was so like perceptual imagery that he was able to build his complex inventions without drawings. Further, he claimed to be able to test his devices in his mind's eye "by having them run for weeks—after which time he would examine them thoroughly for signs of wear."

Although labels lead us to think of the various sensory modes of imagination as though they occur separately, in actuality imagination is polysensory. Albert Einstein, in a famous letter to Jacques Hadamard, described the important role of polysensory (visual and kinesthetic) imagination in his own extremely abstract thinking:

> The words or the language, as they are written and spoken, do not seem to play any role in my mechanism of thought. The psychical entities which seem to serve as elements in thought are certain signs and more or less clear images which can be voluntarily reproduced and combined. . . . The above mentioned elements are, in my case, of visual and some of muscular type. Conventional words or other signs have to be sought for laboriously in a secondary stage, when the above mentioned associative play is sufficiently established and can be reproduced at will.[6]

Although Einstein observed that his polysensory imagination could be directed

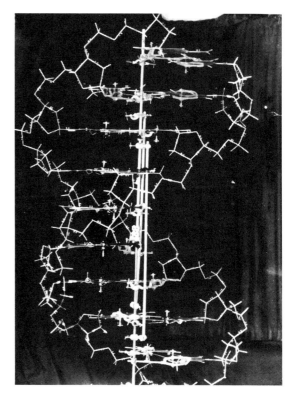

Figure 2-4.

"at will," many important thinkers have obtained imaginative insights more or less spontaneously. For example, the chemist Kekulé came upon one of the most important discoveries of organic chemistry, the structure of the benzene ring, in a dream. Having pondered the problem for some time, he turned his chair to the fire and fell asleep:

> Again the atoms were gamboling before my eyes. . . . My mental eye . . . could now distinguish larger structures . . . all twining and twisting in snakelike motion. But look! What was that? One of the snakes had seized hold of its own tail, and the form whirled mockingly before my eyes. As if by a flash of lightning I awoke.[7]

The spontaneous inner image of the snake biting its own tail suggested to Kekulé that organic compounds, such as benzene, are not open structures but closed rings.

If you identify high intellectual endeavor exclusively with verbal and mathematical symbols, consider the introspections of Tesla, Einstein, and Kekulé with special care. To give another example, the following problem (2-4) is best solved by inner imagery. Has your education prepared you for this kind of problem-solving?

2•4 PAINTED CUBE

Shut your eyes. Think of a wooden cube such as a child's block. It is painted. Now imagine that you take two parallel and vertical cuts through the cube, dividing it into equal thirds. Now take two additional vertical cuts, at 90° to the first ones, dividing the cube into equal ninths. Finally, take two parallel and horizontal cuts through the cube, dividing it into 27 cubes. Now, how many of these small cubes are painted on three sides? On two sides? On one side? How many cubes are unpainted?

Do not be disappointed if you did poorly on this problem. *Mental manipulation of inner imagery improves with practice.*

Experiental exercises for the mind's eye, involving many different kinds of inner imagery, are presented in Section IV, Imagining.

Drawing and Thinking

Very few people possess the acuity of mind's eye that enabled Tesla to design and build complex machinery without drawings. Most visual thinkers clarify and develop their thinking with sketches. Watson, recollecting the thinking that preceded his discovery of the DNA structure, writes that one important idea "came while I was drawing the fused rings of adenine on paper."[4] An example of a chemical diagram drawn by Watson is shown in Figure 2-5.

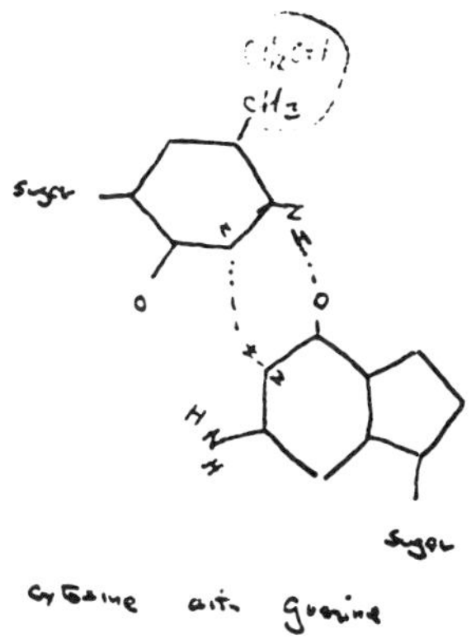

Figure 2-5.

As in Watson's experience, drawing and thinking are frequently so simultaneous that the graphic image appears almost an organic extension of mental processes. Thus Edward Hill likens drawing to a mirror: "A drawing acts as the reflection of the visual mind. On its surface we can probe, test, and develop the workings of our peculiar vision."[8]

Drawing not only helps to bring vague inner images into focus; it also provides a record of the advancing thought stream. Further, drawing provides a capability that memory cannot: the most brilliant imager cannot compare a number of images, side by side in memory, as one can compare a wall of tacked-up idea-sketches.

Two idea-sketches from the notebook of John Houbolt, the engineer who conceived the Lunar Landing Module, are reproduced in Figure 2-6. Houbolt's drawings show two important attributes of graphic ideation. First, the sketches are relatively "rough." They are not intended to impress or even to communicate; instead, they are a kind of graphic "talking to oneself." Second, one sketch is an abstract schematic of the voyage from earth to moon and back; the other is a relatively more concrete side view of the landing module. Idea-sketching, like thinking itself, moves fluidly from the abstract to the concrete.

Drawing to extend one's thinking is frequently confused with drawing to communicate a well-formed idea. *Graphic ideation precedes graphic communication; graphic ideation helps to develop visual ideas worth communicating.* Because thinking flows quickly, graphic ideation is usually freehand, impressionistic, and rapid. Because communication to others demands clarity, graphic communication is necessarily more formal, explicit, and time-consuming. Education that stresses graphic communication and fails to consider graphic ideation can unwittingly hamper visual thinking. The relation of drawing to productive thinking will be treated in Section V, Idea-Sketching.

Some problems are most easily solved by graphic means—for example, problem 2-5.

2.5 WITH ONE LINE

With one continuous line that does not retrace itself, draw the pattern shown in Figure 2-7.

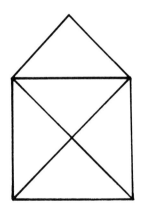

Figure 2-7.

The puzzles used to illustrate various modes of visual thinking in this chapter bear a striking resemblance to visual problems posed in psychological tests of intelligence. In the next chapter, I will briefly review some of these test forms; they will help to shed light on the *operations* of visual thinking.

Answers

2-1: Figure 2-8 shows one of the infinite number of answers to this problem. Can you generate some of the others? For example, what is the "minimum-volume" solution?

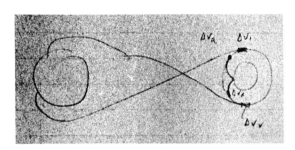

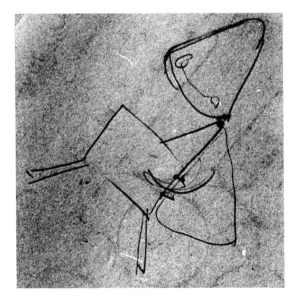

Figure 2-6.

2-2: The fourth card is the correct one.

Figure 2-8.

References

1. Arnheim, R. "Visual Thinking," in *Education of Vision* (edited by G. Kepes). Braziller.
2. Feldman, E. *Art as Image and Idea.* Prentice-Hall.
3. Kaufman, G. *The Book of Modern Puzzles.* Dover.
4. Watson, J. *The Double Helix.* Atheneum.
5. O'Neil, J. *Prodigal Genius: The Life of Nikola Tesla.* McKay (Tartan Books).
6. Einstein, A. Quoted by Hadamard, J., in *The Psychology of Invention in the Mathematical Field.* Princeton University Press.
7. Kekulé, F. von. Quoted by Koestler, A., in *The Act of Creation.* Macmillan.
8. Hill, E. *The Language of Drawing.* Prentice-Hall (Spectrum Books).

Recommended Reading

Rudolf Arnheim's *Visual Thinking* (University of California Press) is a readable and erudite treatment of visual thinking, surely destined to be a classic. If you want to delve deeper into the theory of visual thinking, or if you want to be convinced that visual thinking is the most important kind of thinking, I highly recommend this book.

IMAGES IN ACTION

3

The Operations of Visual Thinking

How do we manipulate visual images in our minds? Psychologists who study and test mental ability have discovered that visual thinking involves a number of visual-spatial operations. The following examples from psychological tests suggest, in a manner that you can directly experience, certain active operations that occur in visual thinking.

The sampling is not exhaustive. Operations that occur unconsciously, such as the fantastic symbolic transformations that occur in dreams, are not represented; nor are the subtler operations of visual synthesis. These and other mental operations will be treated in subsequent chapters. Further, some of the examples exercise several visual-spatial operations at once. In this chapter, experience the problems as an introductory exploration of the kinds of mental operations that do the active work of visual thinking.

Pattern-Seeking

Most of us experience seeing as a passive, "taking-in" process. In fact, *perception is an active, pattern-seeking process that is closely allied to the act of thinking.* The active and constructive nature of visual perception is well illustrated by psychological tests that require the operation of *closure.* In problem 3-1, experience closure in two ways: (1) by "filling in" an incomplete pattern and (2) by finding a desired pattern embedded in distracting surroundings. As you visually seek both patterns, notice that your perception is indeed active.

3•1 FILLING IN[1]

Look at the picture on the left. It is a picture of a violin. What is the figure in the picture on the right?

Figure 3-1.

The violin and camel in Figure 3-1 are incomplete graphic images. So active is the pattern-seeking nature of perception that these partial images are "closed" into meaningful patterns. In the next problem (3-2), perception actively seeks pattern in another way. Once again, experience how actively you visually "close in" on the hidden figure.

3·2 FINDING[2]

The left-hand design in Figure 3-2 is the figure. You are to decide whether or not the figure is concealed in any of the four drawings to the right.

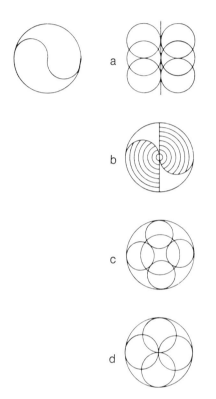

Figure 3-2.

You have "closed" correctly in Figure 3-2 if you found the desired figure embedded in drawings b and d. Now try another kind of visual operation involving pattern perception (3-3). This operation is extremely common and is fundamental to all thinking.

3·3 MATCHING[3]

The first figure in each line of Figure 3-3 is duplicated in one of the five figures that follow. Check the duplicate figure.

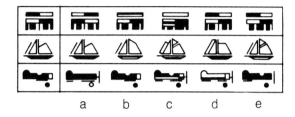

Figure 3-3.

You recognized the patterns correctly if you checked d, a, and b, respectively. When this problem is given in a psychological test, it is called "perceptual speed" and the score is the number of correct matchings completed in a given time. Why does the test emphasize speed? Did you notice that there is a quick and a long way to perform the matching? The long way involves detail-by-detail comparison and perhaps even a bit of talking to oneself. The quick way involves seeing the desired pattern as a whole and matching it without hesitation. The desired operation is the quick way—thus the emphasis on speed in the test.

Computers programmed to do pattern recognition laboriously perform the matching operation in the previous test the long way. Indeed, the computer is a floundering visual thinker, squandering great effort on operations that humans perform almost effortlessly. The following more difficult operation (3-4), essential to the invention of categories, is relatively routine for humans but a tour de force for a well-programmed computer.

3·4 CATEGORIZING[4]

In each of the rows in Figure 3-4, two figures are exactly alike. Check these figures.

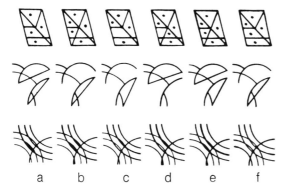

a   b   c   d   e   f

Figure 3-4.

You may rate yourself effective in visual categorization if, in Figure 3-4, you checked b and e, b and f, b and e. We literally invent our world by this visual-spatial operation. On a rudimentary level, *we discover all of the objects in our environment by recognizing common features.* Ask a child how he or she distinguishes cats from dogs. On a more rigorous level, visual categorizing is at the core of much scientific discovering. A skilled eye peering into a microscope operates much as your eye did in the previous example.

The active nature of pattern-seeking is experientially dramatized by tests that involve drawing. As you work the next problem (3-5), realize that your pencil, as it constructs the required image, reflects the constructive activity of eye and mind.

3·5 PATTERN COMPLETION[5]

Complete the patterns in the spaces in Figure 3-5.

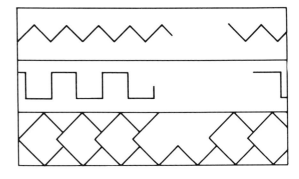

Figure 3-5.

Visual Memory

Ability to retain visual imagery is difficult to measure. One can never be sure that a low test score is the result of poor memory; it could as well be the result of inaccurate perception. Indeed, vigorous perception and faithful re-membering are closely allied. The more actively you perceive the following designs (3-6), the more likely you will be able to reproduce them from memory.

3·6 MEMORY FOR DESIGNS

Inspect the designs in Figure 3-6 for two minutes. Close the book; on separate paper, reproduce them in any order.

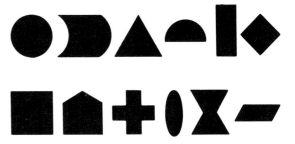

Figure 3-6.

Rotations

So far, you have been actively filling in, finding, matching, categorizing, completing, and remembering fixed images. In the next two test problems, experience the operation of mentally rotating images in space. First, experience rotating a flat image through 180° (3-7).

3•7 INVERSE DRAWING[5]

In Figure 3-7, the top drawing on the left is the top right drawing seen reversed. In the remaining squares, draw the inverse of the drawings on the right.

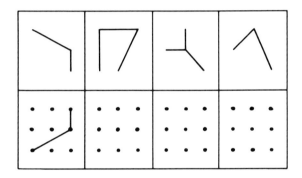

Figure 3-7.

Now mentally rotate the image of a three-dimensional object in space (3-8).

3•8 ROTATING DICE[4]

Examine each pair of dice in Figure 3-8. If, insofar as the dots indicate, the first die of the pair can be turned into the position of the second one, place a check next to the pair.

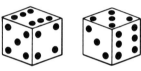

Figure 3-8.

If you checked the second pair of dice, you correctly performed the operation of rotation.

Orthographic Imagination

The rotating-dice problem requires an operation similar to orthographic imagination, which is the ability to imagine how a solid object looks from several directions. This is an alternative operation: either the object is mentally rotated, or the viewpoint is rotated in relation to the object. Orthographic imagination also includes cutting through a solid object and viewing the resulting cross-section (see Chapter 13). Try out your orthographic imagination in problem 3-9.

3·9 FROM ANOTHER VIEWPOINT

In Figure 3-9, the left-hand drawing represents a solid object. One of the drawings in the right-hand column shows the same object in a different position. Circle the number of that drawing.

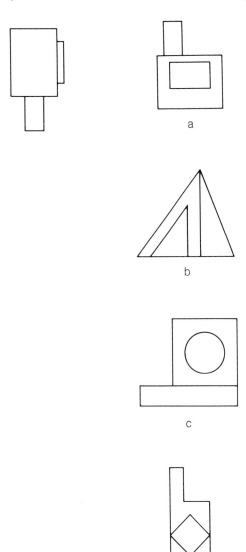

Figure 3-9.

Orthographic imagination is related to, but is not quite the same as, visual-spatial operations that require a strong sense of body orientation. If you were able to see that the correct drawing in Figure 3-9 is c, can you also orient yourself to the following situation? "My house faces the street. If a boy passes by my house walking toward the rising sun, with my house at his right, which direction does my house face?"

Dynamic Structures

In the following visual-spatial operations, the structure of a three-dimensional object is manipulated in some way. In problem 3-10, the object is folded.

3·10 FOLDED PATTERN[6]

The pattern on the top left in Figure 3-10 can be folded to form a three-dimensional object, with the grey showing outside. Which of the four lettered objects is the result?

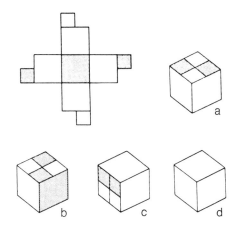

Figure 3-10.

You folded the pattern correctly if you selected object a. Now try *un*folding.

3·11 KNOTS[7]

Which of the drawings of a piece of string in Figure 3-11 would form a knot if the ends were pulled tight?

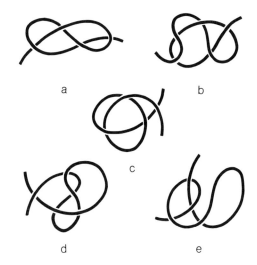

a   b

c

d   e

Figure 3-11.

You correctly performed the required operation if you visualized the loops of string marked b, c, and e as knots.

The last two operations required moving a single configuration in space. Now move several objects in relation to each other. Motion in visual-spatial operations is likely effected by kinesthetic (muscle) imagery. Although problem 3-12 is mechanical in nature, *imagery in three-dimensional motion is important to visual thinking in many fields.*

3·12 PULLEYS[6]

In Figure 3-12, which way (a or b) will pulley "X" turn?

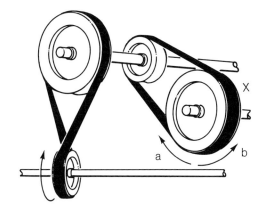

Figure 3-12.

Did you trace the motions of the pulleys with your finger, or feel some sort of inner muscular involvement, as you came to the correct conclusion that pulley "X" goes in direction b? If so, you were experiencing the importance of kinesthetic imagery to active thinking operations.

Visual Reasoning

An alternative way to solve the previous problem of the pulleys, without rotating the pulleys together in imagination, is by visual logic: "This one turns this way, therefore. . . ." A similar kind of visual reasoning underlies problem 3-13.

3·13 SPATIAL ANALOGY[5]

The first three designs on the left in Figure 3-13 are part of an analogy. Which of the designs on the right completes the analogy?

Figure 3-13.

Logical reasoning, on the concrete level of the previous examples, works much the same way in visual thinking as it does in verbal and mathematical thinking. Deductive reasoning, by which the thinker goes from an abstract to a concrete idea, also occurs in all kinds of thinking. But visual deduction, especially of the more profound sort, is difficult to illustrate or describe. The painter, for example, who realizes an abstract idea in a particular composition thinks deductively. "But," as Edward Hill writes in *The Language of Drawing*, "attempting to unravel the workings of this intelligence will baffle the mind entirely. The processes are nearly inscrutable."[8]

Inductive visual reasoning, because it begins with concrete imagery, is easier to illustrate. In problem 3-14, you are asked to induce the abstract principle that relates four sequential images to a fifth image.

If you solved the previous problem (the answers are a and b) rather easily, and are beginning to believe that the operations of visual thinking are necessarily simple, then try problem 3-15. Or better yet, try chess.

3•15 VISUAL INDUCTION II[9]

The two designs in the first row of Figure 3-15 are related to each other and also to one of the designs in the second row. Which of the four lettered designs below fits into the empty space?

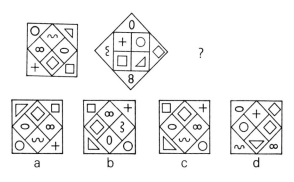

Figure 3-15.

3•14 VISUAL INDUCTION I[6]

On the left of Figure 3-14 are four "problem figures" that are a related series. On the right are five "answer figures," one of which is the fifth figure of this series. For example, the answer figure in the first line is d; we induce that the relating principle is an incremental "falling" of the line to the right. Can you induce the remaining answer figures?

PROBLEM FIGURES

ANSWER FIGURES

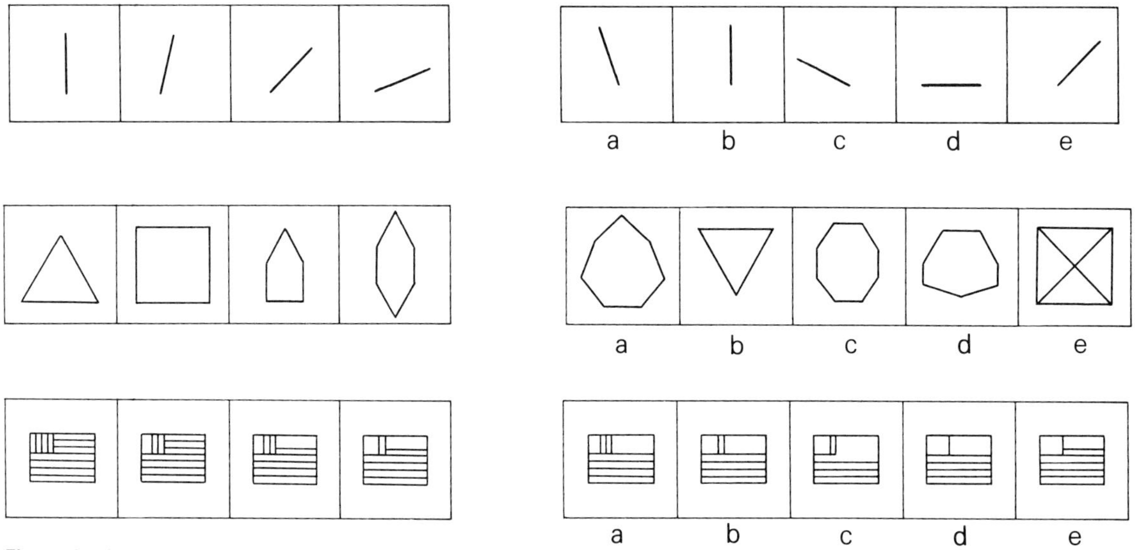

Figure 3-14.

The principle that relates the three designs is sequence: (1) the entire figure rotates 45 degrees; (2) the two sets of symbols move as sets, 90 degrees clockwise and 90 degrees counterclockwise; and (3) inner and outer sets of symbols exchange positions. The related design is therefore c.

Visual Synthesis

As we come to the complex and manifold operations involved in the creative act of visual synthesis, we reach the limits of examples from psychological tests. In one well-known test, shape fragments are put together into wholes—but this is spatial addition, not synthesis. In creative synthesis, the whole is a new identity that is more than the sum of its parts. Another test presents a set of simple geometric shapes that are to be combined, by drawing, into such images as a clown, a microphone, and an ice cream cone. This test, although it provides more latitude, is nevertheless rudimentary. To suggest the complex nature of visual synthesis, we must probe deeper. All of the visual-spatial operations discussed in this chapter, and some additional operations to be introduced later, are involved in the act of creation. But such operations as occur in dreams are closer to the heart of creative visual thinking.

References

1. Thurstone, L., and Jeffrey, T. *Closure Speed (Gestalt Completion)*. Industrial Relations Center.
2. Thurstone, L., and Jeffrey, T. *Closure Flexibility (Concealed Figures)*. Industrial Relations Center.
3. Thurstone, L., and Jeffrey, T. *Perceptual Speed*. Education-Industry Service.
4. Bernard, W., and Leopold, J. *Test Yourself*. Lancer.
5. Smith, I. *Spatial Ability*. Knapp.
6. Bennett, G., Seashore, H., and Wesman, A. *Differential Aptitude Tests: Form L*. The Psychological Corporation.
7. J. T. Bain et al. *Moray House Space Test I*. University of Edinburgh.
8. Hill, E. *The Language of Drawing*. Prentice-Hall (Spectrum Books).
9. Eysenck, H. *Know Your Own I.Q.* Penguin (Pelican Books).

Recommended Reading

If you enjoy performing visual-thinking tests, you will find more of them in books such as those listed above. (Eysenck's paperback is an especially rich source.) Although intended to test inherited ability, visual tests are useful as exercises to awaken and develop resources that education commonly ignores.

AMBIDEXTROUS THINKING

cal center for such symbolically left-handed processes as feeling and imagining, while the left hemisphere serves as the center for such symbolically right-handed processes as verbal reasoning and mathematical description.

Scientific evidence of hemispheric specialization was first observed over a century ago. Patients with brain damage gave the first clues: doctors began to find that those with damage to the left hemisphere were frequently unable to talk, while those with damage to their right hemisphere, though able to talk, suffered impairment to their spatial awareness, memory of faces and places, and artistic and musical ability.

Starting in the 1960's, scientific knowledge about hemispheric specialization was greatly expanded by research with "split-brain" patients whose traumatic epileptic seizures had been relieved by a radical

Left and Right

According to the ancient symbolism of the left and right hands, the left hand represents openness, receptivity, subjectivity, playfulness, feeling, motivation, and sensory and imaginative processes; the right hand represents discipline, logic, objectivity, reason, judgment, knowledge, skill, and language. The symbolic left hand is open to fresh impressions, hunches, and subconscious levels of thinking; the symbolic right hand holds the tools necessary to develop, express, and realize ideas, to bring them into the world of action.

Although the symbolic left and right hands should not be confused with actual left- or right-handedness, the symbolism nevertheless bears some relation to neurological fact. As Figure 4-1 shows, the right hemisphere of the brain crosses over to control the left side of the body, and the left hemisphere crosses over to control the right side. Contemporary neurologists and psychologists have discovered, confirming the insights of the ancients, that *the right hemisphere of the brain serves as the neurologi-*

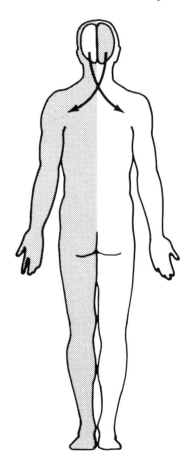

Figure 4-1.

surgery that severed the large bundle of nerve fibers connecting the left and right hemispheres of the brain. Studies with these patients confirmed that the left hemisphere is the primary center for verbal expression, while the right hemisphere is the primary center for comprehending spatial relationships. Electroencephalographic studies have similarly shown that the right hemisphere is relatively less active when a person is trying to solve a verbal or mathematical puzzle, that is, when the left (verbal) hemisphere is at work. When the same person is given a visual puzzle to solve, the left hemisphere relaxes while the right (spatial) hemisphere goes to work.

Not many decades ago, scientists had found very little to justify the existence of the right hemisphere; new research, however, testifies to its important role. In addition to its ability to comprehend spatial relationships, the right hemisphere, while it cannot express itself verbally, can recognize the written word. Other experiments suggest that the right hemisphere has virtually no sense of time, and therefore little sense of cause-effect relationships; instead, it is nonrational, willing to suspend judgment and to think without basis in reason or fact; it is also intuitive and holistic, preferring to perceive and conceive in wholes, simultaneously. The right hemisphere is the sensuous half of the brain, relishing the concrete and the here-and-now; it is also the subjective poet, finding reality in metaphor and perceiving emotional nuances.

By contrast, the left hemisphere not only expresses itself in words but prefers to think "objectively" with symbols and abstract concepts. The left hemisphere has a strong sense of time, and therefore naturally explores sequential, logical, cause-effect relationships. It is rational and therefore seeks reasons, facts, and quantities expressed in numbers.

So far, this chapter has been addressed primarily to your left brain; it is time to switch hemispheres by means of an experience (4-1) that will give you a direct sense of the difference between left- and right-hemisphere consciousness.

4•1 FOOD FOR THOUGHT

The next time you sit down to a meal, experiment with the following two modes of consciousness:

1. As you look at your plate, think about the number of calories in each portion and the percentages of the meal that are protein, carbohydrate, and fat. Analyze and name the constituents of each dish. Look at labels to identify chemical additives that you are consuming; determine whether your meal is well balanced and nutritious. Calculate how much the food cost. Above all, *talk* while you eat! How do you feel as you address your meal primarily with your left hemisphere?

2. Now move your consciousness toward the right hemispheric mode. Pleasurably merge into the experience of savoring your meal: nonjudgmentally attend the taste, texture, and aroma of every bite. Don't talk, listen, or otherwise subject your experience to labeling. Relaxedly, slowly (indeed, without being aware of time) assimilate the meal with all your senses. Be aware of nuances: close your eyes to focus on your nonvisual senses; open your eyes to the pattern and contrasts of the colors on your plate, the plate on the table, the table in its environment. Feel the mood of the place, its spatial quality. With clarity, *be absorbed* in your immediate experience.

3. Alternate between right hemispheric and left hemispheric modes until you can clearly feel the difference between the two, as well as their respective values.

Hemispheric Dominance?

At this point, left-handed readers may well be wondering where they fit into this discussion of left and right hands and hemispheres. With a few exceptions, left-handed and ambidextrous people have the same left-right hemispheric specialization as do right-handers. In other words, the "dominant hand" has little to do with hemispheric specialization for most people. But is there a "dominant hemisphere"?

The next experiment (4-2) demonstrates that one eye (and hence one hemisphere) must dominate vision to prevent a double image.

4.2 DOMINANT EYE

1. Stretch your arm out directly in front of you. *With both eyes open,* point a finger to a distant object.

2. Close one eye; open it, while simultaneously closing the other eye. Alternating left and right eyes, notice that each eye sees a slightly different image. (See how your pointed finger appears to shift off target.)

3. *Close your left eye.* If your finger continues to point toward the object, the eye you are looking through, your right eye, is dominant. If your finger appears to shift away from the object, your left eye is dominant.

To observe that one hemisphere dominates a given function does not lead to the conclusion that one hemisphere dominates in general. For example, until recently, the left hemisphere was called "major" and the right "minor"; this suggestion of hemispheric dominance is falling into disuse as researchers learn that the left and right hemispheres of the brain perform different but equally important functions.

To be sure, genetic, educational, and cultural forces do form what could be termed "hemispheric preferences," even "hemispheric habits." When you look at the drawing in Figure 4-2, what do you see? If you initially saw a duck, you are exhibiting left-hemisphere preference (not surprisingly: you have been reading). If you saw a rabbit, you preferred to process the image initially with your right hemisphere. Since the brain can process only one meaning at a time, hemispheric preference made the first choice. However, now that you have experienced both meanings, you can, of course, choose to see duck or rabbit at will. Similarly, no matter which hand or eye is dominant, or which hemisphere you have learned to

Figure 4-2.

prefer, *you can learn to move your thinking from one hemisphere to another, at will.* In other words, you can *train* your thinking to be ambidextrous.

Bridges Within

The symbolism of left and right hands is sometimes used to classify people into opposing camps, or cultures, much as C. P. Snow once distinguished scientists from literary and artistic humanists. But such classifications can be misleading if they are pressed too far. Truly creative people in every field are ambidextrous in their thinking; that is, they are capable of receiving the inner messages of the left hand and transferring them over to the form-giving, outward-oriented right hand for expression. In Chapter 2, we saw how a number of famous scientists were eminently ambidextrous, gaining spectacular insights from their inner imagery and even dreams. Only later were these discoveries confirmed in the right-handed world of the laboratory. Similarly, a novelist's "sharp ear for dialogue" and ability vividly to imagine characters and settings are gifts brought by the left hand, but to translate these perceptions and imaginings skillfully into words is the work of the right.

In the next experience (4-3), exercise your ability to use your symbolic right and left hands, and to obtain an internal transfer between the two.

4·3 INTERNAL TRANSFER

1. In a pleasant setting, spend at least ten minutes attending your sensations and feelings, and avoiding such symbolically right-handed activities as inner speech and self-evaluation. Begin by closing your eyes and listening to the sounds of your environment. Be aware only of sounds; let all else retreat into the background of your awareness. Also try not to label the sounds (don't think "that's an airplane, that's the wind in the trees"). Labeling is right-handed and restricts full sensory experience. After listening for several minutes to sounds, take a slow stroll, first touching things, then smelling, then seeing. Next, attend your body sensations. Close your eyes and locate various parts of your body; your right big toe, the end of your nose. Be aware of your breathing, without trying to control it (controlling is right-handed). And so on. Finally, attend your emotions. What is your mood right now: tranquil? anxious? euphoric? annoyed? mixed emotions?

2. Transfer some part of the previous experience to your symbolic right hand. As evocatively as possible, verbally describe something of what you have just sensed or felt to a friend, or write about it or draw it. A measure of your ambidexterity in this exercise is your ability to convey your experience to someone else.

That internal transfer from left to right is vital to healthy thinking is an important theme in the work of Abraham Maslow, a pioneer of humanistic psychology.[1] Maslow discerned two levels of creativity in people. Those who make their contributions "by working along with a lot of other people, by standing upon the shoulders of people who have come before them, by being cautious and careful," are capable only of what Maslow calls "secondary creativeness." The person whose creativeness is secondary deals with the outer world logically, objectively, and in orderly fashion, but has lost intimate contact with senses, feelings, and inner fantasy life.

What Maslow calls "primary creativeness" emanates from the unconscious. It is the result of ability "to fantasy, to let loose, to be crazy, privately." Conscious primary creativeness, according to Maslow, "is very probably a heritage of every human being," but most people lose it as they grow up. Most people, that is, whose society demands reality-adjusted thinking only, and whose education has been almost exclusively right-handed.

The most creative thinkers, according to Maslow, are those that have achieved psychological integration. "A truly integrated person can be both secondary and primary; both mature and childish. He can regress and then come back to reality, becoming then more controlled and critical in his responses." *Visual thinkers in every field will be creative only to the extent that they are able to develop this flexibility and integration.*

Symbolically, the left hand has frequently been associated with the impulsive, unconscious, "dark" side of human nature. Ambidextrous thinkers are, of course, actively left-handed; are they consequently open to sinister impulses? Why is the symbolically right-handed person, whom Maslow describes as relying on secondary creativity, afraid of the unconscious? Is there good reason to be fearful? Does evil lurk within, held in check only by the law-abiding discipline of the symbolic right hand?

Opinion on this question is mixed. Freud reached the pessimistic conclusion, along with many Western theologians and philosophers, that human nature is inherently flawed by instinctual drives that can be held in check only by the civilized, law-abiding right hand. Maslow, by contrast, took a more optimistic view: "Complete health," he wrote, "means being available to yourself at all levels."[1] Certainly there is no simple way to reconcile these opposing views. We may well acknowledge that people who have full access to their left hand (who are aware of their senses, open to their feelings, in touch with their unconscious) inevitably experience more of everything—more pleasure, and also more pain. But if the left hand has sinister possibilities, then how should we characterize an overly repressive right hand? In the cartoon by Jules Feiffer (Figure 4-3), notice which hand is holding the gun.

Nietzsche once wrote, "Every extension of knowledge arises from making conscious the unconscious." Ambidextrous thinkers, who are capable of drawing on their primary creativity, necessarily make the unconscious conscious. By contrast, people who are afraid of the unconscious have limited access to this vital source of their being. They are, in other words, blocked from full functioning: fearing themselves, they can utilize only that safe, encultured, law-abiding side which we have called right-handed. *A major purpose of experiences in visual thinking is gently to take the symbolic left hand out of the cast in which society and education have immobilized it, to give it some exercise, and to put it to work in harmony with the right.*

As Maslow suggested, people who are capable only of secondary creativity have little access to primary processes. They stand on other people's shoulders, thinking about the written thoughts of someone else, who in turn was writing about some-

Figure 4-3.

one else's idea, and so on. Verbal thinkers, especially, tend to think in this second-hand way: they skillfully manipulate symbols but rarely make full contact with their own primary resources. Visual thinking, with its symbolically left-handed, primary-process origins, is a vital complement to symbolically right-handed, secondary-process thinking-by-words-and-numbers.

Take perception, for example. What sort of thinking may we expect from thinkers who verbally label their experience before they have time to perceive its richness? Since perception provides vital raw material for thinking, we may expect stereotyped thinking, thinking in which conventionalized labels play a central role. Visual thinking, with its emphasis on fresh and unlabeled perception (see Chapter 8), should have a salutary influence on the thinker who has become over-involved in verbal (or mathematical) labeling. I am not suggesting that we abandon words: much of the material in this book would be extremely difficult to present nonverbally. My point is that *visual and verbal thinking are complementary.*

The same case can be made in connection with sensory imagination. The importance of this primary kind of imagination to thinking in every field cannot be overstressed. Do you suppose, for example, that mathematicians think only in relation to mathematical symbols? An extensive study of creative mathematicians found that almost all of them use imagery, most frequently visual, but sometimes kinetic.[2] Again, I am not attempting to discourage symbolic thinking. By attempting to solve problem 4-4 by *sensory imagination only,* you will see why mathematicians also think in mathematical symbols.

4.4 FIFTY TIMES

Picture a large piece of paper, the thickness of this page. In your imagination, fold it once (now having two layers), fold it once more (now hav-

ing four layers), and continue . . . folding it over upon itself fifty times.

How thick is the fifty-times-folded paper?

Most people cannot sustain this imaginative feat beyond several foldings. Even if they could, the image would be impossible to scan: by mathematical calculation, you will find the fifty-times-folded thickness is in the order of the distance from here to the sun! Most creative mathematicians are visual thinkers, but this example shows why they also use abstract mathematical language. Again, it is not a question of one or the other: sensory imagination and symbolic thinking are complementary, each performing mental functions that the other cannot.

In a similar vein, visual thinking complements abstract-language thinking by its power to make things concrete. The graphic image ■ is clearly more concrete in meaning than the word "square" or the numerical expression "$1/8 \times 1/8$." This is not to say that visual imagery cannot be abstract: visual imagery was central to the extremely abstract thinking of Einstein. Rather, visual imagery can be more concrete than words or numbers; it is in this way that "a picture is worth a thousand words."

Visual and verbal (or mathematical) thinking are also complementary by virtue of differences in structure. Verbal and mathematical symbols are strung together linearly in conventional patterns such as those afforded by grammar. Mentally tracking these linear structures automatically enforces certain thinking operations. As if fingering beads, the thinker follows the verbal or mathematical pattern piece by piece, toward a single end. Visual imagery, by contrast, is holistic, spatial, and instantly capable of all sorts of unconventional transformations and juxtapositions. Thus "the image method remains the method of brilliant discovery," and the verbal-mathematical method, "the way of rationalization and inference."[3]

Computers cannot see or dream, nor can they create: computers are language-bound. Similarly, thinkers who cannot escape the structure of language, who are unaware that thinking can occur in ways having little to do with language, are often utilizing only that small part of their brain that is indeed like a computer. "Language," writes Arthur Koestler, "can become a screen which stands between the thinker and reality. This is the reason that true creativity often starts where language ends."[4] Creative thinkers do what the computer cannot. They abandon language when occasion demands, and enter into other modes of thought. Specifically, creative thinkers are ambidextrous: they use the symbolic left hand as well as the right, the right brain as well as the left. Learning to think visually is vital to this integrated kind of mental activity.

References

1. Maslow, A. "Emotional Blocks to Creativity," in *A Source Book for Creative Thinking* (edited by S. Parnes and H. Harding). Scribner's.
2. Hadamard, J. *The Psychology of Invention in the Mathematical Field.* Princeton University Press.
3. Bartlett, F. *Remembering.* Cambridge University Press.
4. Koestler, A. *The Act of Creation.* Macmillan.

Recommended Reading

Blakemore, C. *Mechanics of Mind.* Cambridge University Press.

Bogen, J. E. "Some Educational Aspects of Hemispheric Specialization," in *UCLA Educator,* 17 (1975).

Bruner, J. *On Knowing: Essays for the Left Hand.* Harvard University Press.

Buzan, T. *Use Both Sides of Your Brain,* E. P. Dutton.

Edwards, B. *Drawing on the Right Side of the Brain.* J. P. Tarcher.

Gazzaniga, M. "The Split Brain in Man," in *Perception: Mechanisms and Models* (edited by R. L. Held). W. H. Freeman.

Jaynes, J. *The Origin of Consciousness in the Breakdown of the Bicameral Mind.* Houghton Mifflin.

Masters, R., and Houston, J. *Listening to the Body.* Delacorte Press.

Ornstein, R. *The Psychology of Consciousness.* Penguin Books.

Russell, P. *The Brain Book.* Hawthorn.

Savary, L., and Ehlen-Miller, M. *Mindways: A Guide for Exploring Your Mind.* Harper & Row.

Sperry, R. W. "Hemispheric Disconnection and Unity in Conscious Awareness," in *American Psychologist,* 23 (1968).

PREPARATIONS

The following two chapters are concerned with two important, though commonly neglected, preparations for visual thinking. The first describes materials and environmental conditions that are especially conducive to visual thinking. The second introduces the value of an inner state of relaxed attention.

MATERIALS AND ENVIRONMENT

5

Visual-Thinking Materials

Although visual ideas can be captured with whatever is at hand—a stubby pencil on the back of an old envelope, a finger in the moist surface of a sandy beach—visual thinking benefits from an appropriate choice of materials. Not all visualization materials are well suited to exploring and recording ideas. Materials that involve difficult techniques, for instance, will absorb energy and divert attention away from thinking. Time-consuming techniques also impede rapid ideation, since ideas frequently come more quickly than they can be recorded. Frustration with an unwieldy material can block a train of thought or be reflected directly in diminished quality of thinking. The best materials for visual thinking are direct, quick, and easy to use.

The following list is intended to aid the beginner, but the experienced visual thinker or professional can profit from it by taking a moment to evaluate current practices with the intent of trying something new. If you are accustomed to sketching on 8½'' × 11'' paper, try 18'' × 24''; better yet, try a roll of newsprint. Get a better marker; explore the clarification power of color. *Breaking up old sketching*

habits can prove to be a powerful renewal strategy, for a new pattern of idea expression can readily open the gates through which all ideas must pass.

Introductory List of Materials

Markers:
1 nylon-tip pen (black)
1 ballpoint pen (black, medium point)
5 Magic Marker or Eagle Prismacolor felt-tip markers (assorted colors). Note: when buying felt-tip markers, consider warm grays (Magic Markers #2 and #4 or Prismacolor 8971 and 8974, for example) for quickly rendered shading effects.
3 Prismacolor pencils (assorted colors). When buying colored pencils, avoid the hard kind; Prismacolor's color range, intensity, and texture are difficult to beat.
1 Conté stick (black #3)
Paper:
1 large newsprint pad (approximately 18'' × 24'')
1 tracing pad (approximately 14'' × 17'', with one sheet of grid paper)
Eraser:
1 Pink Pearl
Tools:
1 mat knife
1 inexpensive pair of scissors
1 wooden ruler with brass insert (24'')

Are there special instructions for the use of these materials? "Restrictive directions as to how to hold tools and step-by-step instructions for working can make for self-conscious and mechanical procedures, inhibiting rather than furthering the direct and vigorous expression of native capacities."[1]

5•1 EXPERIMENTATION

Play freely with the materials that you have purchased so far. Try different markers and paper combinations. Experiment with line quality, such as light strokes and bold strokes. Which markers are best for filling in solid tones? Graduated tones? Which marker-paper combination gives you the most pleasure?

Because visualization materials can profoundly influence the way you think, you may want to experiment with other materials and their combinations. Well-supplied art-materials stores carry markers of many kinds. Try carbon, charcoal, pastel, and lithograph pencils. You may also enjoy using pens like the Rapidograph or Koh-i-noor. A word of caution: drawings made with markers such as pastel and charcoal, even when sprayed with fixative, are messy to store.

From the wide variety of papers available, choose the less expensive, especially if you are a beginner. *Costly paper tends to inhibit thinking.* Newsprint paper, convenient to buy by the pad, is cheaper by the ream (500 sheets) or roll. Newspaper print rooms will sometimes give you newsprint in the form of unused "roll-ends." Manila wrapping paper, purchased by the roll, provides another inexpensive drawing surface. Keep in mind, though, that inexpensive pulp papers discolor and become brittle with age. Rag-content papers should be used for permanent records. Although more expensive, they are more stable chemically and have the additional virtues of whiteness, erasability, and hard finish.

In addition to the drawing tools listed, you may want to purchase a plastic triangle, a circle guide, and an ellipse guide—to be used sparingly. Visual ideas are generally best recorded freehand, with a minimum of tools.

Mockup materials have not been so well developed by manufacturers as drawing materials. Clay, the traditional sketch material of the sculptor, has many disadvantages to weigh against its basic advantage of malleability. Clay's soft plasticity tends to limit, and even to define, the kinds of forms that can be visualized; it directs ideation to surface considerations; it is heavy, messy, and time-consuming to use. Styrofoam is an important alternative to clay: it is relatively stiff, can be easily formed into a hollow structure, and can be glued. Geometric forms are usually best developed with sheet materials. Sheet thermoplastic (such as Plexiglas or styrene) is easily cut by scoring and breaking, can be formed with heat, and is quickly joined with plastic solvent. Foamcore, a laminated paper-and-Styrofoam sandwich, is a stiff and light sheet material that cuts like butter; it is especially well suited for large mockups.

Of course, all visualization materials can be used in combination. A free sense of experimentation is the key. However, do not become so enamored of materials and techniques that you forget that they are only a means to more effective visual thinking. Pleasurable play with materials is a very tempting alternative to productive mental activity.

Optical Equipment

In addition to the inexpensive materials described so far, consider using various kinds of optical equipment as tools for visual thinking. Cameras, useful for making "record shots," can also be used creatively. For example, a quickly developed Polaroid transparency of a small sketch can be projected on a wall-size piece of paper, where details can be developed in larger scale. Photographs and drawings can be combined. Slide projectors can be used to stimulate visual thinking—for example, by means of multi-screen projections of visual material relevant (and perhaps irrelevant) to the task at hand. Opaque projectors offer additional possibilities. For that matter, any optical device that extends the power of the eye—even simple magnifying and reducing lenses—can be a tool for visual thinking.

Environment for Visual Thinking

As a skilled surgeon would not consider operating in a poorly illuminated and ill-organized room, so you should carefully attend to the environment in which you plan to perform *mental* operations. Your work area should be well illuminated, preferably with natural north light and without shadow or glare. The drawing surface should be large and adjustable in height and angle. An additional table should be available for three-dimensional work; spilled glue and knife marks soon spoil a drawing surface. Organized storage should be arranged close to each work area to minimize distracting clutter. Chairs and stools should provide back support in the working position. To alleviate back tension, and also to provide for the important element of change, a stand-up, vertical drawing surface should be available: a blackboard, easel, or wall-mounted roll of paper. A large tack-space is needed for displaying current idea sketches. Although admittedly an affront to those who associate productive work with open eyes and erect posture, *a quiet place where you can relax and turn your thoughts inward—or stop thinking entirely—is essential: a reclining chair, a couch, even a relaxing bath!*

Also consider the subjective nature of your environment:

> Some of the stimuli with which certain great thinkers sought to surround themselves are curious and even bizarre; yet their presence seems to have been strangely necessary to creative thought. . . . Dr. Johnson needed to have a purring cat, orange peel, and plenty of tea to drink. . . . Zola pulled down the blinds at midday because he found more stimulus for his thought in artificial light. Carlyle was forever trying to construct a soundproof room, while Proust achieved one. Schiller seems to have depended on the smell of decomposing apples which he habitually kept concealed in his desk.[2]

You will be much more productive working in surroundings that are aesthetically stimulating and emotionally comfortable for you than in surroundings that rub you the wrong way.

What about several people thinking together? Opinion is divided on the subject of group-think. Some believe that superior ideas are invariably the gift of the individual thinking alone. Others hold that group interaction stimulates individual thinking, and even that an openly interactive group can form a "group-mind" with breadth of information and powers of association unattainable in a single mind. Although these opinions are not mutually exclusive, one cannot entertain the possibility of group-think, especially visual group-think, without coming directly upon environmental considerations. It can be easily demonstrated, for example, that five people sitting in a straight line cannot interact verbally as well as can five people sitting in a circle. When the interchange is visual as well as verbal, environmental considerations become even more critical.

I am not going to try to describe an optimal environment for visual group-think, but I shall mention a few of the problems that must be overcome. Clearly, an interactive group needs to be able to work over a shared visual image, suggesting modifications and changes, making erasures, and so on. A group working at a single blackboard can do this. But at a blackboard, what happens to A's idea that B erased in order to present an alternative? One problem is the need for a memory device. Also, how is the group going to generate a variety of visual concepts and then, together, compare and evaluate them? To my knowledge, no one has yet devised a workable system to facilitate visual group-think. Such a system would be extremely useful, however, for thinking about complex, interdisciplinary problems—problems of the urban environment, for example.

The Graphic Computer

The graphic computer is rapidly becoming an extremely important tool for visual thinking. An interactive graphic computer, such as the one shown in Figure 5-1, allows visual thinkers to manipulate graphic imagery in space and time, to have access to a vast visual computer memory, to decrease their involvement (by means of the computer's memory function) in routine visualization tasks, and most important, to handle more complexity faster. When a system for visual group-think is devised, it will very likely incorporate a graphic computer system that will permit colleagues to

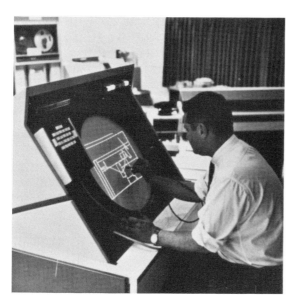

Figure 5-1.

interact, even though they may be separated by thousands of miles. At this writing, the exciting potential of the graphic computer has not been fully realized. But what is in hypothetical or experimental form now will likely be in prevalent use in the near future.

References

1. Mendelowitz, D. *Drawing: A Study Guide.* Holt, Rinehart & Winston.
2. McKellar, P. *Imagination and Thinking.* Basic Books.

Recommended Reading

For a description and discussion of an environment constructed at Stanford for facilitating inner imagery, see my chapter entitled "The Imaginarium: An Environment and Program for Opening the Mind's Eye," in *Visual Learning, Thinking, and Communication,* edited by B. S. Randhawa and W. E. Coffman (Academic Press, 1978).

RELAXED ATTENTION

6

The Paradox of Ho-Hum and Aha!

Relaxation involves loosening up, letting go, and finally—ho-hum—going to sleep. Attention involves focusing energy, finding excitement in discovery—aha!—and being very much awake. Ho-hum and aha!—what can these seemingly opposite modes of consciousness have in common? Together, how are they related to visual thinking?

Relaxation and attention are two sides of the same paradoxical coin. The first tenet of skill in any field is relaxation: the skilled always "make it seem easy." The second tenet is complete attention: expert practitioners invariably "give their all." Indeed, relaxation and attention are mutually supportive. By relaxing irrelevant tension, the individual releases full energy and attention to the task at hand. Watch any masterful performance—a rhythmic golf swing, a breathtaking ballet leap, a virtuoso violin solo—and you will see the importance of relaxed attention.

And so it is with thinking, our highest skill. Relaxation is important to thinking generally, because we think with our whole being, our body as well as our brain: "Nothing," writes Harold Rugg, "is

more basic than the role of the body. We not only move with it, we think with it, feel with it, imagine with it."[1] Overly tense muscles divert attention, restrict circulation of blood, waste energy, stress the nervous system: uptight body, uptight thoughts. Be reminded, however, that a totally relaxed person cannot think at all, even though awake. Physiologists have shown that some muscular tension is needed to generate and attend mental processes. Some tension, but not too much: relaxed attention.

The importance of relaxed attention to creative thinking is well known. After intensive conscious preparation, the creative thinker commonly lets the problem "incubate" subconsciously: "I will regularly work on a problem late into the evening and until I am tired. The moment my head touches the pillow I fall asleep with the problem unsolved."[1] After a period of relaxed incubation, which can take place in the shower or on a peaceful walk as well as during sleep, attention is not uncommonly riveted by the "aha!" of sudden discovery. "Frequently I will awaken four or five hours later . . . with a new assembly of the material."[1] While subconscious incubation requires relaxation, a sudden flash of insight requires attention, or it is lost. Again, relaxed attention.

Memory, as Aldous Huxley reminds us, operates in much the same fashion:

> Everyone is familiar with the experience of forgetting a name, straining to capture it and ignominiously failing. Then, if one is wise, one will stop trying to remember and allow the mind to sink into a condition of alert passivity; the chances are that the name will come bobbing up into consciousness of its own accord. Memory works best, it would seem, when the mind is in a state of dynamic relaxation.[2]

The ability to relax attentively is especially important to visual thinking. *Excessive eye tension interferes with seeing; directed imagination is enhanced by a disciplined form of letting go; visual ideas flow most freely onto paper when the marker is held and moved with graceful ease.* Even more than most human skills, seeing, imagining, and drawing require relaxed attention.

Optimal Tonus

Relaxed attention occurs when the relative balance of relaxation and tension brought to a task is appropriate. Bernard Gunther, in *Sense Relaxation*, calls this balance "optimal tonus."[3] Edmund Jacobsen, in *You Must Relax*, calls it "differential relaxation."[4] Both concepts describe the human organism adjusting dynamically and economically to the task at hand, but never pushing or straining unnecessarily.

Jacobsen describes differential relaxation in terms of primary and secondary activities. He defines primary activities as those essential to the desired behavior. For reading a book such as this one, these activities include contracting certain muscles to maintain your posture and moving your eyes to follow the words. Secondary activities are those that detract from primary activities. For example, if you hear a noise in the other room while you are reading this book, you may glance up for a moment or even turn in that direction. In differential relaxation, the person differentiates between primary and secondary activities, relaxing primary activities that are "unnecessarily intense for their purpose" and carrying "relaxation of secondary activities to the extreme point, since these are generally useless."

Gunther emphasizes that optimal tonus, or differential relaxation, is "not letting go completely. Sleepy-sagginess-collapse is the opposite pole to hypertension." Excessive relaxation immerses one in lethargy or sleep.

A uniquely human way to let go of excessive tension and return to optimal tension is to laugh: laughter is usually a sign that an individual or a group is pleasantly relaxed. Scornful laughter, however, can trigger tension. To understand the cause of tension, to get to its very roots, we must probe more deeply.

Causes of Excessive Tension

When we tense our muscles, we do so to make a response. But why do we overtense our muscles? What are the causes of excessive tension?

By far the most fundamental cause is fear. Fearful or insecure people tense up because they believe that they will soon face a real or imagined attack or catastrophe. At work, they overreact and burn energy needlessly, or do not act at all—in each instance, to avoid failure. At home in bed, they fuss and worry; their bodies tense, they cannot go to sleep.

Unable to relax, fearful people also find it difficult to maintain attention. Every distraction is interpreted as a potential threat or taken as an opportunity for relief. Easily diverted, they become prone to the conflicting mental agenda and immobile tension that characterize the indecisive.

Excessive tension has physical as well as psychological causes. When energy level and muscular structure are inadequate to the response demanded of them, excessive tension results. When relatively unmuscular intellectuals, for example, let their highly active minds place unsupportable demands on their ill-conditioned bodies, their bodies "say no" by tightening their muscles.

"The person who is well and doesn't worry, who works consistently but without overeffort, who has sense enough to rest when he becomes tired, will never have to learn special techniques for relaxing. But how rare is that person!"[5] Indeed, so prevalent is tension that large industries cater to letting go: alcohol, drugs, entertainment, health clubs, and vacation resorts. And when tension becomes unbearable, we go to the last resorts, psychologists and psychiatrists.

The Carrot or the Stick?

At this point, you may be comparing what you have read so far with your own experience and beginning to form questions such as "Isn't relaxation something you do when you get home and take off your shoes?" Or "Isn't it possible that relaxation is okay for some, and not for others?" Or "What about people who seem to be at their best in an atmosphere that literally crackles with tension, tight deadlines, and the threat of impending disaster?" Questions such as these reveal an underlying concern for motivation, and indeed, survival.

Remember the "carrot" and "stick" theories of motivation? If you wonder how you could perform without adrenaline-inspiring deadlines, you may be confessing your dependence on the stick (fear) as a motivating force. Most of us have been brought up and educated, at least in part, by the "stick." Whether self-inflicted or wielded by others, the stick *does* inspire action. So what is wrong with it? *Stick-inspired action is unpleasant, sporadic, and most important, conforming. Fear instills obedience and thwarts risk-taking.* If you aspire to independent, creative thinking, you would be well advised to reflect on the role that fear of the stick plays in your motivational life.

Is the "carrot" a valid alternative to the stick? The external carrot, in the form of money, prestige, power and so on, is usually the stick in disguise. Although apparently motivated by reward, the external carrot-seeker is more often moved by fear of failure to obtain *enough carrot*. What remains, then, is the internal carrot that draws us to our full potential, that attracts us to perform at our best for the sheer pleasure of doing things well. *Truly creative people are most attracted by the nourishment afforded by their internal carrot and relatively less motivated by external rewards and fear.*

Relaxed attention is a fear-reducing, creativity-inducing strategy based on the internal-carrot theory of motivation. Relax, stick lovers: you have nothing to lose but fear itself.

Breaking the Cycle

Jacobsen, in *You Must Relax,* persuasively maintains that physical-relaxation techniques provide an excellent way to break the cycle of fear, worry, and tension.[4] Observing that physiological tests "indicate that when you imagine or recall about anything, you tense muscles somewhere, as if you were actually looking or speaking or doing something," he counsels us to learn to observe the muscular sensations that accompany our negative thoughts. By relaxing these muscular tensions, Jacobsen claims, we can diminish our negative psychological state. A totally relaxed person, Jacobsen has demonstrated, cannot entertain any kind of thinking at all, worrisome or constructive.

Can the psychosomatic cycle of fear and tension be broken by deep physical relaxation? Possibly not without expert and sustained training in relaxation techniques, and in extreme cases, probably not without additional expert psychological counsel. However, the experiential exercises that follow are nonchemical, free, available at any time, and relatively easy to learn. Try them. They are excellent preparation for visual thinking and may even have additional benefits.

Letting Go

Relaxation experts generally agree that the first step in eliminating excessive muscular tension is to become aware that you are tense. The next step is to realize that *you* are responsible for your excessive tension. The final step is to learn to *let go of tension voluntarily.* Ways of letting go take two forms: dynamic and passive. Dynamic letting go involves activity; passive letting go involves lying down and going limp. Most relaxation techniques combine both forms, as does the simple technique of stretching (6-1).

6·1 STRETCH

---

1. Close your eyes and sit quietly for several minutes. Allow your attention to systematically explore the muscle sensations of your body: your face muscles, neck muscles, shoulders, and so on down. Where are you excessively tense right now?

2. Now, stand up and stretch—slowly, gracefully, and luxuriantly, like a cat. As you do, inhale deeply and feel the tension in your body.

3. With a generous sigh, exhale, sit down, and relax. As you do, feel the tension letting go. Sustain this passive sensation for several minutes.

Letting go of neck and shoulder tension (6-2) is a special problem for people who "work with their heads." The human neck was evolved for the flexible side-to-side and up-and-down head movements required for hunting and survival. Holding the heavy human head over a desk for long periods while looking rigidly straight ahead at paperwork is a comparatively recent behavior that places an extremely unnatural demand on neck and shoulder muscles. These areas should be relaxed periodically, and always just before intensive visual/mental activity.

6•2 RELAX NECK AND SHOULDERS

1. Very slowly bend your head forward three times, backward three times, and to each side three times. Then slowly circle your head through the same movements, clockwise then reverse, three times each. Go slowly and gently; most civilized necks are stiff. You will likely have to do this exercise several weeks before you can do it comfortably.

2. Pull your shoulders as far forward as you can, then as far up, as far back, as far down. Repeat three times.

3. With the fingers of both hands, massage the nape of your neck (near the back of the skull). Better yet, invite a friend to massage your neck and shoulder muscles gently with long strokes down along the back of the neck and shoulders. Then return the favor.

4. Take a deep breath, and, with a sigh, let go excess neck and shoulder tension . . . more . . . more . . . passively let go.

6•3 RELAX ARMS AND HANDS

1. Sit or stand erect. Let your arms and hands hang loosely at your sides, like wet spaghetti.

2. As loosely as possible, shake your right hand. Extend this action to your forearm, then your entire arm. Let your arm rise over your head, shaking the entire limb loosely and vigorously.

3. Stop and compare the feeling of your right arm with that of your left.

4. Repeat with your left arm.

The importance of relaxed vision to visual thinking should go without saying. Tired, strained eyes interfere not only with visual thinking but with efficient mental functioning generally. Here is a basic way to relax your eyes (6-4).

6•4 PALMING

1. Precede this passive form of eye relaxation by gently massaging your temples and the nape of your neck and by blinking to lubricate your eyes.

2. "In palming, the eyes are closed and covered with the palms of the hands. To avoid exerting any pressure on the eyeballs (which should never be pressed, massaged, rubbed, or otherwise handled) the lower part of the palms should rest upon the cheekbones, the fingers upon the forehead. . . . When the eyes are closed and all light has been excluded by the hands, people with relaxed organs of vision see their sense-field uniformly filled with blackness."[2] Put your elbows on a desk, or on your knees, so that you can hold your head comfortably on your palms for several minutes.

3. If you see any imagery at all in your mind's eye, your eye muscles are not fully relaxed. Turn your imagination to a pleasurable scene involving black: a furry black cat resting on a large black velvet pillow, or the night sky. You may have to palm your eyes several times a day, and for a period of time, before your inner field of vision takes on the deep and rich blackness that characterizes complete eye relaxation.

In the next method of eye relaxation (6-5), you make existing tensions greater and then release them. By bringing muscles that cause excessive tension into awareness, you can then consciously let them go.

6•5 FACIAL AND EYE MUSCLES

1. Wrinkle your forehead upward. Become aware of the muscles that are controlling the tension, and progressively let them go.

2. Frown tightly and let go.

3. Shut your eyes tightly and let go.

4. With eyes open, look to the far left, become aware of the muscles involved, and let go. Repeat looking right, up, and down.

5. Conscious of the tensions involved, look forward at a distant, then a near object. Now let your vision unfocus and your eyes relax.

6. Close your eyes. In your mind's eye, imagine a bird flying from tree to tree, then perching quietly; a ball rolling along the ground, then coming to a stop; a rocket being launched, then disappearing into the blue sky; a slow, then very fast Ping-Pong match; a rabbit hopping. . . .

Unfortunately, many methods for tension release are too conspicuous or time-consuming for people to use at work, where they need to relax the most. One exception is a relaxation technique advocated for centuries: deep breathing. We all must breathe, whether busy or not. You can quickly demonstrate to yourself the effectiveness of deep breathing as a method to release tension (and also to increase energy). Many deep-breathing techniques have been proposed. The one suggested here (6-6) is simple and reinforces the desired state by a kind of autosuggestion.

6•6 DEEP BREATHING

Slowly and easily take a deep breath, filling the bottom of your lungs as well as the top. As you breathe in, whisper the syllable "re." Pause for a moment, then breathe out, whispering the syllable "lax." Don't force the air in and out of your lungs; let it flow slowly and naturally: re-e-e-e (pause) la-a-a-a-ax (pause).

Deep muscle relaxation (6-7) prepares you to sleep—"perchance to dream"—or, if mental alertness is retained, to imagine more vivid and spontaneous visual fantasies than can usually be obtained with normal muscle tone (see Chapters 15 and 17).

6•7 DEEP MUSCLE RELAXATION

1. Lie down in a comfortable and quiet place.

2. Systematically (a) tense a specific muscle group (listed in step 3 below), (b) study the feeling of tension, and (c) relax, studying the feeling of letting go. If possible, step 3 should be read to the person who is relaxing, the reader giving the relaxer ample time (and occasional reminders) to become aware of the feeling of tension and of letting go in each muscle group. The slash (/) signifies a pause.

3. Clench fists / Flex wrists (back of hand toward forearm) / Hands to shoulders, flex biceps / Shrug shoulders (touch ears) / Wrinkle forehead up / Frown / Close eyes tightly / Push tongue against roof of mouth / Press lips together / Push head back / Push head forward (chin buried in chest) / Arch back / Take deep breath, hold it, exhale / Suck stomach way in / Tense stomach muscles (as if someone were going to hit) / Tense buttocks / Lift legs, tensing thighs / Point toes toward face, tensing calves / Curl toes down, tensing arches / Repeat each activity above, letting go tension in each muscle group even more. Feel the peaceful, positive feeling that accompanies deep relaxation.

Devoting Attention

Letting go is often such an attractive experience that another reminder may be in order: no activity, mental or physical, can be performed in a state of total relaxation. *Some tension is essential to attention.* The goal of relaxed attention is to let go of chronic, excessive, or irrelevant tension, so that energy may be directed appropriately, freely, and fully. Devoting attention is the same as focusing energy. The vehicle for transmitting human energy is muscular tension.

There are many varieties of attention, however, some undesirable. In the military, attention is an order: "a-ten-*shyun!*" Too often the classroom also takes on a military air: several dozen people, despite large differences in personal interest, are forced to "pay attention," equally and together. When this external demand for attention becomes internalized, we force ourselves to pay attention. Externally or internally demanded, *forced attention* usually occurs for brief moments only and must continually be reinforced.

Paying attention because you should or ought to is clearly less pleasant, and less effective, than devoting attention because you want to. When you attend because you want to, you are not easily diverted. *Immersed attention* is natural absorption in developing an idea, contemplating an object, or enjoying an event. Watch a child pleasurably engrossed in stacking blocks to obtain a clear image of immersed attention.

Immersed attention should not be confused, however, with *passive attention*, which is being easily absorbed, willy-nilly, in whatever comes. The passively attentive child who "seems to belong less to himself than to every object which happens to catch his notice" presents a formidable challenge to a teacher. Passive attention "never is overcome in some people, whose work, to the end of life, gets done in the interstices of their mind-wandering."[6]

Preattention is another natural form of attention. Absorbed in thought, for example, you suddenly realize that you have somehow negotiated your automobile through miles of turns and traffic without conscious awareness: you have been preattending the driving task. Preattention is comparable to an automatic pilot that attends routine events but cannot cope with the unusual. Should a highway emergency occur while you are preattending, you must come to full attention to cope with it.

William James described another mode of consciousness, calling it *dispersed attention:* "Most of us probably fall several times a day into a fit somewhat like this: The eyes fixed on vacancy, the sounds of the world mix into confused unity. . . . The foreground of consciousness is filled, if by anything, by a sort of solemn sense of surrender to the passing of time."[6] Unlike preattention, dispersed attention is not accompanied by another train of thought. Dispersed attention rests the human organism; it is a natural function of the attend-withdraw, tidal character of consciousness.

Of the kinds of attention discussed so far, immersed attention would at first seem best suited to visual thinking. What could be better than being able to "lose oneself," to become wholly immersed in what one is doing? Emphatically better is a quality of attention to which sense of self is not lost and consciousness is not taken over entirely by what one is attending. I will call this kind of attention *voluntary attention.*

When you attend voluntarily you are able to change the focus of your attention quickly, at will. To do this, your consciousness cannot be wholly immersed; you must be sufficiently self-aware to be able to decide.

The ability to direct attention voluntarily, and to sustain attention, is central to human freedom. "The essential achievement of [free will]," wrote William James, "is to attend a difficult object and hold it fast before the mind." An idea "held steadily before the mind until it fills the mind" automatically steers behavior; when ideas "do not result in action, it will be seen in every such case, without exception, that it is because other ideas rob them of their impulsive power."[6]

Like the art of relaxation, skill in voluntary attention can be learned. The first principle to learn is that *you can fully attend only one thing, or related group of things, at a time* (6-8). True, you can preattend one thing (of a routine nature) and attend another. But try to attend fully two unrelated conversations at a time, and you will find that you can do so only by alternating your attention between the two. You will also find that your attention naturally favors the conversation that most interests you, which introduces a second principle of voluntary attention: *find interest in what you are attending, or your attention will wander, become divided, or have to be forced* (6-9).

6•8 ATTENTION IS UNDIVIDED

"For a brief period pay attention to some visual object—for example, a chair. As you look at it, notice how it clarifies itself by dimming out the space and objects around it. Then turn to some other visual object and observe how this, in turn, begins to have a different background."[7] Notice also that what was once a sharply focused figure merges into a relatively undifferentiated and unfocused background when attention is shifted. Perception naturally seeks one meaningful pattern at a time—in the terms of Gestalt psychology, one "figure-ground relationship" (see Chapter 10).

6•9 ATTENTION FOLLOWS INTEREST

Again, allow an object in your immediate environment to become a figure against a ground. This time, however, also allow your *feelings* about the object to come into a clear figure-ground relationship. Become aware of whether you like or dislike the object. If your feelings are neutral, be aware that objects or ideas that "leave you cold" are not easy to attend. An object that disinterests you is far more difficult to attend than one that you like or dislike. Attention follows feelings of interest, positive or negative.

Many people confuse staring with visual attentiveness. Asked to look at something, they stare at it. Staring, however, is not only inattentive; it is also bad vision. The fovea, a small patch of sharp focus on the retina of the eye, must scan the attended object freely in order to obatin a complete image. Thus, the third principle of voluntary attention is that *attention is dynamic* (6-10). Whenever mind and eye become immobile, attention diminishes and vision blurs.

6•10 ATTENTION IS DYNAMIC

"Stare fixedly at any shape, trying to grasp precisely this shape by itself and nothing else. You will observe that soon it becomes unclear and you want to let your attention wander. On the other hand, if you let your gaze play around the shape, always returning to it in the varying backgrounds, the shape will be unified in these successive differentiations, will become clearer, and will be seen better."[7]

"If we wish to keep [our attention] upon one and the same object, we must seek constantly to find out something new about it," observed William James. This last principle, that *voluntary attention is an act of continual discovery*, implies curiosity and more (6-11). Continued James: "Attention is easier the richer in acquisitions and the fresher and more original the mind. . . . And intellect unfurnished with materials, stagnant, unoriginal, will hardly be likely to consider any subject long."[6]

6•11 ATTENTION IS CONTINUAL AHA!

Select an object that pleases you. See how long you can find something new about it. View it from many angles. Explore it with all your senses. Imagine how it is made. Contemplate the origins of its constituent materials, the kinds of skilled people involved in its creation. Compare its qualities with like qualities in other objects (for example, if it is red, compare it with other red objects). And so on . . .

Clearing the Ground for Relaxed Attention

By now, I hope the relationship between relaxation and attention is becoming intellectually and experientially evident. By relaxation, you let go inappropriate muscular tensions that divert energy from what you are doing; by attention, you direct and devote your energy, freely and dynamically, to discovering more and more about a single object, idea, or activity that interests you. Old habits, however, may initially make the task of maintaining a state of relaxed attention difficult. Excessive tension reappears; the mind wanders. Exercise 6-12 will help to clear the ground of consciousness so that the physical, mental, and emotional awareness inherent in relaxed attention can be maintained for longer periods of time.

6•12 CLEARING THE GROUND

1. Sit comfortably erect, hands on your thighs, eyes closed. Follow your thoughts for a minute or so, without judging them.

2. Now gradually begin to attend your breathing. "With your eyes closed, experience the air moving in and out of your nostrils. Become aware of the whole cycle of inhalation–pause–inhalation–pause."[3]

3. Now count your breathing. Count 1 on inhalation, count 2 on exhalation, and so on up to 10, and then begin again.

4. Now allow your attention to disperse. Gently relax your eyes. Let your mind go blank: it will, when you are sufficiently relaxed. If thoughts come, don't attend them with a sharp rebuff; follow them, slowly. Gradually, enter into silence.

This last experience, an exercise in relaxed attention, is also a form of meditation. The experience in exercise 6-11, continually finding something new about an object, is a form of contemplation. Picasso said this about contemplation and meditation:

For me, creation first starts by contemplation, and I need long, idle hours of meditation. It is then that I work most. I look at flies, at flowers, at leaves and trees around me. I let my mind drift at ease, just like a boat in the current. Sooner or later, it is caught by something. It gets precise. It takes shape . . . my next painting motif is decided.[8]

Although frequently taught to athletes and performing artists, the art of relaxed attention is rarely even mentioned to those who would develop skill in thinking. Indeed, contemporary education commonly inculcates fear, tension, and forced attention, to the detriment of thinking. Relaxed attention can be educated, and should be. Negative thinking habits are reinforced by education for thinking that does not treat relaxed attention as an essential preparation.

References

1. Rugg, H. *Imagination.* Harper & Row.
2. Huxley, A. *The Art of Seeing.* Harper & Row.
3. Gunther, B. *Sense Relaxation.* Macmillan (Collier Books).
4. Jacobsen, E. *You Must Relax.* McGraw-Hill.
5. Rathbone, J. *Relaxation.* Columbia University Press.
6. James, W. *Psychology: The Briefer Course.* Harper & Row (Harper Torchbooks).
7. Perls, F., Hefferline, R., and Goodman, P. *Gestalt Therapy.* Julian Press.
8. Picasso, P. Quoted by Gauthier, S., in *Look Magazine,* December 10, 1968.

Recommended Reading

An effective method that combines relaxation, autosuggestion, and visualization is described in J. H. Schultz and W. Luthe's *Autogenic Training* (Grune & Stratton). Bernard Gunther, in *What to Do Till the Messiah Comes* (Macmillan–Collier Books), presents a beautifully illustrated guide to other ways of obtaining the graceful state of relaxed attention, including massage. John White and James Fadiman's *Relax: How You Can Feel Better, Reduce Stress, and Overcome Tension* (Dell), a superb selection of three dozen articles, introduces you to a variety of other approaches to relaxation. And Herbert Benson's *The Relaxation Response* (William Morrow) provides an easy-to-learn meditation technique along with medical research findings that validate this mental approach to a state of relaxed attention.

SEEING

The following seven chapters are intended to help you to integrate thinking and seeing, and to vitalize seeing generally. Seeing is treated not as a passive process but as a creative, active art that can be learned.

EXTERNALIZED THINKING

7

Thinking and Seeing

Traditionally, thinking has been considered a symbolic activity quite separate from seeing. According to this view, seeing is sensory information-gathering; the higher mental activity of thinking is verbal or mathematical information-processing. While thinking, one can look out the window or examine one's fingernails—further testimony to a distinction between thinking and seeing.

But is such a distinction psychologically sound? As shown in Chapter 3, many visual operations (such as categorizing and reasoning) have counterparts in symbolic thinking operations. In this chapter, consider the possibility that thinking and seeing can function together. Consider the sculptor who thinks in clay, the chemist who thinks by manipulating three-dimensional molecular models, or the designer who thinks by assembling and rearranging cardboard mockups. All are thinking by seeing, touching, and moving materials, by externalizing their mental processes in physical objects. Many contemporary thinkers, in science and engineering as well as art and design, respect the fertility of this venerable form of visual thought.

Do not be confused by the similarity between externalized visual thinking and the expression of visual thought. A chemist who is advancing his or her thinking by playing with a molecular model is not involved in the same process as a chemist who is using a molecular model to communicate a fully formed idea to another person. *Externalized thinking involves actively manipulating an actual structure, much as one would manipulate that structure mentally.*

Externalized thinking is best accomplished with materials that are easy to form and reform. The sculptor's Styrofoam, the chemist's snap-together elements, the designer's Foamcore and tape all have the virtue of being easily manipulated spatially, much as symbols and images are moved and modified internally in mental space. Materials used to communicate a visual idea that is already formed need not be as flexible.

Externalized thinking has several strategic advantages. First, direct sensory involvement with materials provides sensory nourishment—literally "food for thought." Second, thinking by manipulating an actual structure permits serendipity—the happy accident, the unexpected discovery. Third, thinking in the direct context of sight, touch, and motion engenders a sense of immediacy, actuality, and action. Fourth, the externalized thought structure provides an object for critical contemplation as well as a visible form that can be shared with a colleague or even mutually formulated. Finally, the intense sensory/spatial involvement of externalized thinking fires up the right hemisphere of the brain; it is a marvelous antidote for thinking that has become locked up in words and symbols.

Begin the first experience in the section on seeing (7-1) by externalizing your thought processes—literally by seeing what you think. Link perception, thinking, and action as closely together as you possibly can. Cut; fold; touch; test; hold the pieces together in a new way. Externalize your thinking, as if the process were described accurately by one word, "perceive-think-act."

7•1 TOWER OF PULP

With two sheets of newsprint and 24 inches of Scotch tape, construct the tallest tower that you can in 30 minutes. You may cut, fold, or form these materials any way you like. Other challenges, using the same materials: the longest bridge, the largest enclosed volume (open or closed), or the strongest 12-inch-high support structure (add increments of weight to test).

Many sports require thinking and doing in the immediate context of seeing. The fascination of visual puzzles can also be attributed to the way they involve us in this ancient and fundamental thinking process. As you work the following puzzles, you may experience the pleasurable sensation of "aha!" just before you see the solution. W. J. J. Gordon, in *Synectics,* calls the "pleasurable mental excitement" that often precedes the solution of artistic or technical problems the "hedonic response."[1] Beware: the hedonic response is addicting!

The first puzzle (7-2) is an example of a two-dimensional experience in exter-nalized thinking invented many years ago by a Chinese puzzlemaker.

7•2 TANGRAMS[2]

Begin by dissecting a cardboard square into seven pieces, as shown in Figure 7-1.

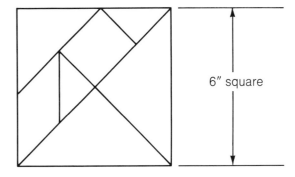

6" square

Figure 7-1.

The invariable rule in solving a tangram puzzle is that you must use all seven tangram pieces to form a given silhouette. Figure 7-2 shows a frisky dog. As you structure the tangram pieces to form this silhouette, be aware that your ability to move the pieces around makes the solution easier.

Figure 7-2.

A "tangram paradox" is shown in Figure 7-3. Although the same seven pieces were used to construct each silhouette, one of the "men" has a foot and the other hasn't. Try to solve this paradox mentally first, and then by moving the pieces. Which is easier in this case: thinking by imagining, or thinking by seeing? The likely answer is a bit of each.

The following three-dimensional thinking-by-seeing puzzle (7-3) can be homemade or can likely be purchased at your local game store. As you work this puzzle, *be aware of the way you search for a solution.* Do you move the elements about by an intuitive form of trial and error? Do you use some form of visual logic? Or do you quietly look at the elements until you imagine a solution? More power to you if you can solve Soma cube puzzles entirely in your mind! But if you cannot, remember the value of externalizing your thinking when you are solving real-life problems.

7·3 SOMA CUBE

The Soma cube was invented by the Danish poet-scientist Piet Hein, who discovered that when three or four cubes of the same size are combined into all possible irregular configurations within a 3 × 3 × 3 matrix (by joining their faces), the combined cubes can be fitted together into a larger 3 × 3 × 3 cube. Assemble a Soma cube by first cementing 27 children's blocks, or equivalent-sized cubes, into the seven Soma pieces shown in Figure 7-4.

When the cement is dry, begin simply by assembling two Soma pieces into the step form shown in Figure 7-5. Then assemble all pieces into a single cube (there are several ways to do this). Next, see if you can construct the configurations shown in Figure 7-5, each with all seven pieces. Then invent your own configurations.

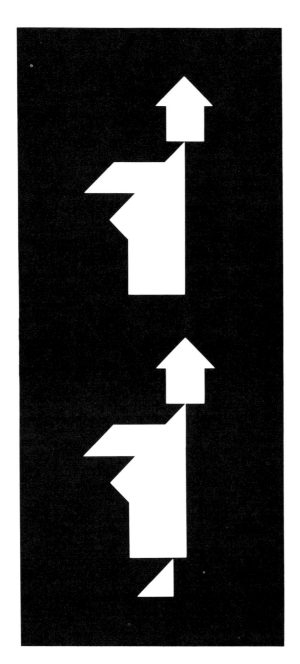

Figure 7-3.

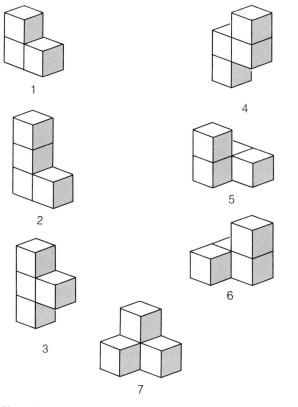

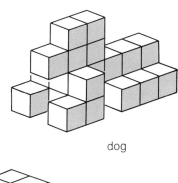

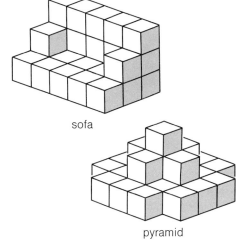

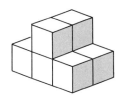

Figure 7-4.

dog

sofa

pyramid

step form made from
two soma pieces

Figure 7-5.

If you enjoy visual puzzles and games, you can readily acquire an extensive collection of interlocking bent-wire puzzles, Chinese wooden puzzles, jigsaw puzzles, and a score of games that involve externalized thinking. For advanced challenge, try the puzzle of four multicolored blocks called Instant Insanity, or the games of three-dimensional tick-tack-toe and chess.

In Section V, externalized thinking will be discussed again in terms of manipulating two-dimensional diagrams and three-dimensional mockups to solve problems that, like the Tower of Pulp and unlike most visual puzzles, have no single right answer. In the meantime, you will have many opportunities to experience the phenomena of seeing more fully. As you will soon see, seeing can be enhanced even when vision is 20/20.

Recommended Reading

Edward de Bono's *The Five-Day Course in Thinking* (Basic Books) affords fascinating challenges in externalized thinking—bottle and knife puzzles and paper cutout games—along with an excellent commentary that leads one to examine how he or she thinks. Martin Gardner's *The Scientific American Book of Mathematical Puzzles and Diversions* (Simon & Schuster) also includes numerous brain teasers and topological curiosities that exercise skill in externalized thinking.

References

1. Gordon, W. *Synectics*. Harper & Row.
2. Read, R. *Tangrams: 334 Puzzles*. Dover.

RECENTERING

Creative Seeing

"The experienced microscopist," wrote Aldous Huxley, "will see certain details on a slide; the novice will fail to see them. Walking through a wood, a city dweller will be blind to a multitude of things which the trained naturalist will see without difficulty. At sea, the sailor will detect distant objects which, for the landsman, are simply not there at all."[1] Why do two companions, one more knowledgeable than the other, see the same world differently? Assuming that both have normal vision, both are sensing, on the retina, essentially the same patterns of light reflected from the environment. But seeing is more than sensing. Seeing requires matching an incoming sensation with a visual memory. The knowledgeable observer sees more than a less knowledgeable companion because he or she has a richer stock of memories with which to match incoming visual sensations.

The knowledgeable observer may not see creatively, however. "Observation," noted Pablo Picasso, "is the most vital part of my life, but not any sort of observation."[2] What sort of observation enabled

Picasso to revolutionize art or Sir Alexander Fleming to discover penicillin "accidentally"? Each was clearly knowledgeable in his field; each was motivated by the tremendous curiosity essential to vital seeing. Equally knowledgeable and curious observers, however, consistently fail to make creative observations. Individuals like Picasso and Fleming create new knowledge by observing the familiar in an unfamiliar way. *Creative seeing occurs when vision is flexibly recentered to a fresh viewpoint by the exercise of creative imagination.*

Imagination and Seeing

William James observed that while "part of what we perceive comes through our senses from the object before us, another part (and it may be the larger part) always comes out of our mind."[3] I have expanded this notion diagrammatically in Figure 8-1, substituting "imagination" for what James called "mind."

On the left in the diagram is the object perceived, in this instance an unclad woman. In the center of the diagram is a built-in "perceptual filter" that mixes incoming sensations with varying degrees of imagination. On the right are the different kinds of perception that result from perceptual filtering.

The top arrow in the diagram represents an *inner image* which is all imagination and no sensation (and therefore not really a sense perception). (Inner, or "mind's eye" imagery is discussed in Section IV, Imagining). The next arrow down represents a hallucination; although triggered by a sensation of the outer environment, hallucination is mostly a product of imagination. A hallucinator looking at the woman pictured might well see the Queen of England and be totally unaware that a large amount of imagination has filtered into perception.

The third arrow down represents projection or stereotyping; in this type of perceptual filtering, the intrusion of imagination is less: unlike the hallucinator, the person who is projecting is not "seeing things." Nevertheless, projection is

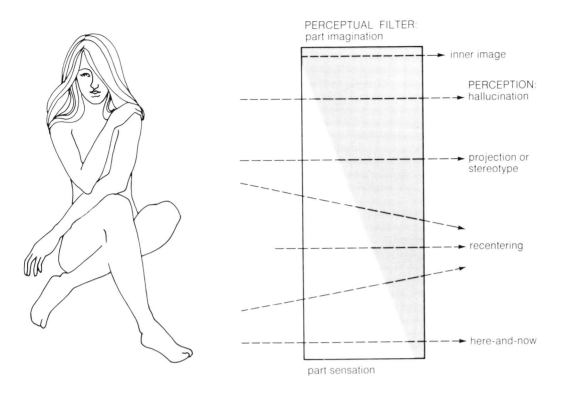

PERCEPTUAL FILTER:
part imagination

inner image

PERCEPTION:
hallucination

projection or
stereotype

recentering

here-and-now

part sensation

Figure 8-1.

strongly influenced by imagination. People who are projecting might like or dislike the lady in the diagram because she evokes memories of a friend, a relative, or even something they like or dislike about themselves. Another instance of projection is seeing faces or animals in cloud forms. The person who perceives stereotypes (for example, that every long-haired man is a Communist) also unconsciously mixes much imagination with sensation. Although projection can be creative (see Da Vinci's Device in Chapter 10), stereotyped seeing is inflexible (once a Commie, always a Commie) and the diametric opposite of creative seeing.

The last arrow in the diagram represents here-and-now perception, which as you will experience in exercise 8-1, is relatively unfiltered by imagination. In this perceptual mode, the perceiver is most open to seeing "what is" and least influenced by imaginative intrusions such as occur in projection and stereotyping. Consequently, recentering into the here-and-now is very important to creative seeing.

Creative seeing involves using imagination to recenter viewpoint; it is the ability to change from one imaginative filter to another. The recentering perceiver might, for example, see the naked woman as would a sculptor (perhaps assessing the formal quality of her pose), then as would an advocate of women's liberation (she is being exploited), then as would the woman herself (I feel a bit chilly), and so on.

Since all perception involves some degree of imagination, we all see imaginatively—in the broad sense of the word. We do not all see creatively, however. The key concept is flexibility. *People who can flexibly use their imagination to recenter their viewpoint see creatively.* People who cannot budge their imaginations to see alternative viewpoints, by contrast, experience only a one-sided, stereotyped vision of reality.

Stereotyped Vision

As we mature, James said, "Most of us grow more and more enslaved to the stock conceptions with which we have once become familiar, and less and less capable of assimilating impressions in any but the old ways."[3] Why? Laziness and appetite for simplicity are inherent in our visual makeup. To help us avoid perceptual chaos, nature blessed us with a "selective mechanism" that is utilitarian, focusing our eyes only on what we need, or need to avoid. When, for example, we want to sit down, we quickly spot a chair and sit. Unless we are at that moment visually active and curious, we will likely not be able to describe this chair an hour later, because we did not really see it. Instead, we saw a "perceptual concept," a stereotyped chair. Our eyes, given voice, might well defend their lazy utilitarianism by saying: "We found you a place to sit, what more do you want?" Breaking away from this natural tendency toward stereotyped vision requires effort.

Visual stereotypes are also socially conditioned. Inherently social, we naturally want to share the visual values of our parents and neighbors. And our parents and neighbors want us to share their values—even by threat of coercion. In our society, prejudice based on skin color is an example of fear-induced visual stereotyping, as is puritanical horror-fascination with nudity. During war, every society propagandizes a dehumanized visual image of the enemy; without such gross stereotyping, war would be intolerable. Fear-induced stereotyped vision, often incurred at an early age, is painfully experienced in neurotic and psychotic behavior. The paranoid individual, for example, projectively sees threat where threat does not exist. Of course, not all socially conditioned vision is fear-induced: the Bantu tribesman positively enjoys the beauty he finds in elongated earlobes. Most of the pleasure of culture, and of cultural comparison, comes from essentially the same enculturation as does the fear-laden stereotype.

To judge whether your own vision is stereotyped, be aware of your emotions with regard to "unacceptable" images. For example, be aware of your feelings when your clothing is somehow conspicuous. If you are afraid to experiment with your clothing, and tend to conform closely to current fashion, your clothing preferences are likely stereotyped at least partially by fear. More important, ask how far you could depart from the visual norm of fashion without having real reason to fear losing friends, losing your job, or even being committed to a mental institution. Social coercion patterns perception more powerfully than we usually realize.

Healthy Perception Is Flexible

Ernest Schachtel observes that children initially pass through a period that seems "open, inexhaustible, exciting, full of wondrous and adventurous possibilities, not to be described by any label." However, with maturity, we see "in most men a slackening of curiosity, fascination, play, exploration, excitement, enthusiasm: the 'open' world has now turned into a variety . . . of objects-of-use to which certain adaptive responses are given." Thus many adults enter into a womblike "cultural cocoon" in which "all objects are reduced to and exhausted by the labels and reactions the culture provides them."[4]

But is the cultural cocoon as sane as its occupants commonly believe? R. D. Laing, in *The Politics of Experience*, ruefully observes: "Society highly values its normal men. It educates children to lose themselves and become absurd, and thus to be normal. Normal men have killed perhaps 100,000,000 of their fellow normal men in the last fifty years."[5] In this light, should we not ask ourselves what is sane and healthy perception?

Frank Barron, who has studied the psychological health of many highly creative people, suggests an answer to this question:

> When an individual thinks in ways that are customarily tabooed, his fellows may regard him as mentally unbalanced. In my

view this kind of imbalance is more likely to be healthy than unhealthy. The truly creative individual stands ready to abandon old classifications and to acknowledge that life, particularly his own unique life, is rich with new possibilities.[6]

Healthy perception is not stuck in a cocoon of cultural conditioning; it is open, flexible, and alive.

Recentering

What is required to recenter vision away from stereotypes, toward healthful flexibility and openness? A slight recentering of vision can be realized at little risk. But to sustain perceptions substantially different from those of others can be a frightening and lonely experience. A major recentering is undertaken only by those who so value what they see outside the cultural cocoon and are so pained by their attempts to conform inside that no alternative to a new vision of reality is viable. Many people seek communal support to sustain radically recentered vision, as did the early Christians. Subcultures, however, soon develop rigid stereotypes of their own. The ability to recenter perception freely, in the long run, depends on the courage and vitality of the individual.

Recentering vision is fundamentally an experience in unlearning. For most people, breaking lazy, category-hardened, fear-induced habits of seeing is an educational task of considerable magnitude. As art educator Edward Hill observes: "Only the extremely exceptional student comes to this discipline unburdened with patent vision. The instructor must unwind a whole circuit of conditioned responses and conventional orientation which seem to deny perception, at least as an active force."[7] R. D. Laing defines the depth of the problem: "Our capacity to see, hear, touch, taste, and smell is so shrouded . . . that an intensive discipline of unlearning is necessary for anyone before one can begin to experience the world afresh, with innocence, truth and love."[5] Even the revolutionary artist Paul Cézanne reported

"waiting for nature to free his eyes from their camera habits."[8]

Knowing that you are in the company of the needy many, and even of the needy great, may help you accept a fundamental condition for unlearning stereotyped vision: you must welcome the insecurity, the adventure, and eventually the wisdom of courting the unknown. Now explore some of the ways to expand and vitalize seeing by recentering.

Here and Now

Recentering, as the term suggests, invigorates vision by moving perception away from its usual viewpoint to a new center where even familiar things are seen differently. Traveling to foreign places is a well-known way of recentering. From the vantage point of an unknown culture, we can more easily see our enculturated selves in a new light.

But people who constantly "travel in their minds," imagining events that are remote in time and space, direct their perceptions away from the only actual reality, the reality of here and now. Energy devoted to pondering what was, or speculating on what might be, cannot be devoted to experiencing the excitement and wonder of what is. The following experience (8-1) recenters awareness toward immediate and direct sensory experience.

8·1 "FEELING THE ACTUAL"

Frederick Perls, founder of Gestalt therapy, introduced this experience by first warning that it "does not imply a constant state of pop-eyed alertness. This would indicate chronic apprehensiveness, which usually rests on a misapprehension of reality." Instead Perls encourages you "to let go . . . and bask in animal comfort" as does a "household tabby."[9] So sit down, relax, and follow these deceptively simple instructions. Try for a few minutes to make up sentences stating what you are at this moment aware of. Begin each sentence with the word "now" or "at this moment" or "here and now." For example: At this moment I hear the song of a bird. Now I am aware of my heart beating. Here and now I feel . . . and so on.

Gently avoid judging what you experience; also note when you involuntarily "travel" imaginatively to "there-and-then."

If you have difficulty recentering your perception into the actuality of the "here-and-now," you may want to try subsequent exercises in Dr. Perls's book *Gestalt Therapy,* intended to help you identify personal "resistors" to awareness of what is.

Making the Familiar Strange

W. J. J. Gordon, in *Synectics,* discusses two basic perceptual modes involved in problem definition: "making the strange familiar" and "making the familiar strange."[10] Our natural tendency, when first faced with a strange situation, is to analyze it, to reduce it to categories, in short, to make the strange familiar by fitting it into an accepted pattern. While this reductive tendency is economic (the human organism naturally seeks the least possible effort), making the strange familiar as quickly as possible leads readily to stereotyped thinking. Consequently, Gordon recommends making the familiar strange: "Basic novelty demands a fresh viewpoint, a new way of looking at the problem. Most problems are not new. The challenge is to view the problem in a new way."

Topsy-turvies (8-2) are conscious attempts to make the familiar strange, actually or imaginatively. An actual topsy-turvy, for example, is literally looking at the world upside-down.

8•2 TOPSY-TURVIES

Ross Parmenter, in *The Awakened Eye,* suggests that you "stand with your back to what you want to observe, legs apart. Then bend over at the waist and, with your head up-side down, look back between your legs at the scene you want to contemplate."[11] Our normal processes of perception are somewhat upset by this maneuver: usual ways of judging distance are made uncertain and everyday associations are reversed. Viewing conversation upside-down can be especially startling; the lower lip takes the place of the upper and appears to be incredibly mobile. With habitual associations diminished, colors also seem more vivid and contrasts of light and dark more intense.

Other examples of actual topsy-turvies are looking at the world by means of distorted reflections (such as are found in polished metal bowls and glassware) and reversing habitual behaviors (for example, eat dinner for breakfast, wear your shoes on opposite feet).

Or try imagined topsy-turvies. Imagine your room with the colors switched around: the floor color on the ceiling, the color of your chair on the wall, and so on. Reverse functions: wear dishes, eat from your hat, go to sleep in a large head of lettuce. Allow fish to fly. You have the idea; now invent your own.

Role playing, in which you play the role of another being or object, has a similar recentering effect. Role playing can be imaginative (ask yourself how a particular situation would appear from a mosquito's viewpoint) or actual (switch roles with an opponent during an argument).

Relabeling

Labels play a surprisingly important role in focusing our perceptions. James observed that children who are asked to see how many features they can point out on a stuffed bird "readily name the features they know already, such as . . . tail, bill, feet. But they may look for hours without distinguishing nostrils, claws, scales, etc., until their attention is called to these details; thereafter, however, they see them every time. In short, the only things which we commonly see . . . are those which have been labeled for us, and then stamped in our mind. If we lost our stock of labels we should be intellectually lost in the world."[3]

Schachtel, however, notes the inherent danger of labels: "The name, in giving us the illusion of knowing the object designated by it, makes us quite inert and unwilling to look anew at the now supposedly familiar object from a different perspective."[4]

Since perception is object-oriented, one way to recenter perception is to abandon object labels and to relabel the environment according to another method of classification (8-3).

8•3 "REDITURE"

Instead of labeling your perceptions according to the usual object categories, label them according to qualitative categories such as color. Instead of seeing groupings of furniture in a room, for instance, see and group first all things that are red, then all things that are yellow. In other words, look for the "rediture" instead of the "furniture." Recenter again by other relabelings: look only for the "cubiture" (all things cubic), the "rounditure," the "smoothiture," and so on. Notice how a familiar room becomes new again: colors become brighter and richer; patterns, shapes, and textures suddenly emerge from the shadows of familiarity. Realize also how often you have allowed the "veil of words" to obscure and stereotype your vision.

A well-known creativity test asks you to list as many alternative uses as you can for a common object, such as a brick. The creative person who is capable of recentering makes a long and diverse list. The person whose perception is rigidly centered on the usual construction function of a brick makes an impoverished list. (This latter tendency to sort objects into indelibly labeled containers is called "functional fixedness.") In the next experience (8-4), take pleasure in recentering your perceptions of function, in relabeling the object with flexible ease: a rose is a paintbrush—and also a cork!

8•4 A ROSE IS A CORK

With Scotch tape on newsprint, assemble parts cut from picture magazines into new composite images of your own creation. Every cut-out in the composite image should have at least two identities: the old one and the new one. For example, the eye in a taped-together composite face might also be a marshmallow, a wheel, or a flower. In addition to finding pleasure in transforming the functional identity of an object, also enjoy putting the object in an unfamiliar, and even shocking, context. Changing the surroundings in which the object is seen can markedly recenter the way it is seen, as suggested by Claes Oldenburg in Figure 8-2.

Figure 8-2.

Ross Parmenter, pointing out that the simile is a kind of relabeling that not only enlivens vision but makes you "feel the magic of what is described,"[11] suggests a visual game based on the simile (8-5).

8•5 VISUAL SIMILES

Viewing an object, ask yourself "What does it recall?" For example, the author Colette described a fire on a hearth as a "glittering bouquet." Make sure that your simile makes a substantial imaginative leap, as from fire to flowers. A lake should not recall another lake, or even a rain puddle, but perhaps a jigsaw puzzle piece, a heat mirage, a fun-house mirror.

When you break away from "objective" reality to the more subjective level of the visual simile, you are experiencing the associative flexibility inherent in the creative act. The strict object-ivity so actively fostered by contemporary education blocks our natural proclivity to express ourselves creatively with the same kind of symbolic transformations that occur in dreams. Visual similes and metaphors help to express the poet within each of us who would glimpse the profound reality that lies beyond labels.

Unlabeling

Semanticists tell us that "the word is not the thing." Relabeling is one way to use words to recenter vision out of restrictive habits of lazy labeling. Another excellent way to recenter seeing is to abandon words altogether, to *un*label.

In the following unlabeling experiences, recenter the way you perceive people. Schachtel observes that "most people, most of the time, see other people as objects-of-use. . . . The perceiver's senses are not directed toward and receptive to the other person as a human being."

8•6 CEREMONIAL LABEL BURNING

Listen to the way you introduce yourself, or a friend, to other people. Introductions tend to be dehumanizing labelings.

Obtain a large gummed label and list on it every term that can be used to describe you (such as Hubert, son, skinny, talkative, lawyer, sincere, funny . . .). Get some help if necessary; a complete human label lists at least 30 names and adjectives. Wear the label for a while, on your forehead perhaps. Then burn it, ceremoniously.

8•7 NONVERBAL COMMUNICATION

Center your perceptions on the way people (including you) express themselves nonverbally. Attend what is "said" by eyes, eyebrows, hands, posture, and tone of voice—by clothes, personal environment, and symbols of status. Don't label these nonverbal perceptions with words. Instead, closely attend the feelings that these often unwitting communications evoke in you.

8•8 BEYOND LABELS

Sit opposite another person at a distance comfortable to both of you so that you have an easy view of the other's face. Without ever talking, each simply experience the other's face. Don't stare: staring, as discussed in Chapter 6, defeats good vision. Visual attention is best sustained by moving the eyes as interest dictates. Again, don't internalize your experience verbally. Go beyond labels to attend the feelings evoked by this human encounter. Sustain this experience for at least 30 minutes to obtain its full impact and value.

References

1. Huxley, A. *The Art of Seeing.* Harper & Row.
2. Picasso, P. Quoted by Gauthier, S., in *Look Magazine,* December 10, 1968.

3. James, W. *Psychology: The Briefer Course.* Harper & Row (Harper Torchbooks).
4. Schachtel, E. *Metamorphosis.* Basic Books.
5. Laing, R. *The Politics of Experience.* Ballantine.
6. Barron, F. "The Psychology of Imagination," in *A Source Book for Creative Thinking* (edited by S. Parnes and H. Harding). Scribner's.
7. Hill, E. *The Language of Drawing.* Prentice-Hall (Spectrum Books).
8. Cézanne, P. Quoted by Rugg, H., in *Imagination.* Harper & Row.
9. Perls, F., Hefferline, R., and Goodman, P. *Gestalt Therapy.* Julian Press.
10. Gordon, W. *Synectics.* Harper & Row.
11. Parmenter, R. *The Awakened Eye.* Wesleyan University Press.

Recommended Reading

Max Wertheimer, in *Productive Thinking* (Harper & Row) describes instances of productive recentering in the thinking of such people as Einstein, Galileo, and Gauss, and leads you to appreciate the power of recentering for your own thinking. Edward de Bono's *New Think* (Basic Books) does much the same thing; de Bono uses the term "lateral thinking" to describe what I have called recentering. E. T. Hall's *The Silent Language* (Doubleday) and J. Fast's *Body Language* (Evans) will help you to recenter your attention away from words and labels and toward the revelations of nonverbal communication. Jim Adams's entertaining book *Conceptual Blockbusting* (W. H. Freeman) describes many kinds of blocks that inhibit perceiving and thinking, and suggests what to do to remove them.

More specific to creative seeing, George Nelson's *How to See: Visual Adventures in a World God Never Made* (Little, Brown) takes you on an eye-opening tour of your man-made environment, Edi Lanners's *Illusions* (Holt, Rinehart & Winston) treats you to a compendium of eye-tricking optical illusions, and Richard Gregory's *The Intelligent Eye* (McGraw-Hill) enables you to experience how you acquire perceptual skills and how these skills are an integral part of the way you think.

To bring all this into balance, see Frederick Franck's *The Zen of Seeing* (Knopf), an intimate view of visual experience that joins the artistic with the spiritual.

SEEING BY DRAWING

Unblocking the Natural Impulse to Draw

Specialized thinking that equates drawing with art keeps many people, including many contemporary curriculum planners, from realizing that visual education need not be education for a life in art. Unlearn the stereotype that places drawing in the category of Art, capital A! Drawing, most of all, stimulates seeing.* It is an inducement to stop labeling and to look. And no more habitual disclaimers about lack of artistic talent: almost everyone learns to read and write in our society; almost everyone can also learn to draw. To unblock your impulse to draw, and thereby refresh your vision, heed Kimon Nicolaïdes as he relabels the act of drawing: "It has nothing to do with artifice or technique. It has nothing to do with aesthetics or conception. It has only to do with the act of correct observation, and by that I mean a physical contact with all sorts of objects through all the senses."[1]

In this and the remaining chapters of this section, *you will learn ways to invigorate the way you see by drawing*. In this chapter you will begin with relaxed doodling that evolves to scribble-drawing of likenesses. In Chapter 10, you will seek the bold, overall pattern of visual images. In Chapter 11, you will recenter your vision into an analytical mode to explore the richness of detail that is embedded in the larger visual pattern. In Chapter 12, you will combine pattern-seeking and analytical ways of seeing to record proportional relationships. In Chapter 13, you will use drawing to investigate the visual cues that enable you to comprehend solid forms and the relationship between forms in space.

Relaxed Eye, Free Hand

Seeing and drawing, like all human skills, are best accomplished in a state of relaxed attention. Before beginning this next experience, return to Chapter 6, and relax your muscles of vision and drawing to a state of "optimal tonus." Review also the exercises in attention. Seeing by drawing involves attending two images—the image of the object in view and the image on the paper. Begin simply (9-1), by attending only one of these images, the image that you are drawing.

9•1   FREE DOODLING

In a playful and relaxed spirit, draw long, sweeping lines with each of your markers. In rhythm with the natural sweep of your hand and arm, vary drawing pressure from extremely light (just grazing the paper) to very heavy. Fill in areas; dot; texture; pattern. All the while, simply enjoy seeing what you are causing to happen on the paper. Return to that time in your early childhood when drawing was pleasurable doodling that didn't have to look like anything.

To make the marker go where you want it to go requires eye-hand coordination. Strengthen this visual-motor cooperation by making up doodling games such as those in exercise 9-2.

*Photography also enlivens vision. If possible, supplement the drawing sequences that follow with related work in photography.

9.2 DISCIPLINED DOODLING

1. At random, pepper a sheet of newsprint with a couple of dozen dots. Connect the dots with a pattern of horizontal, diagonal, and vertical lines. Draw each line freehand with a single, decisive stroke. As you draw, keep your eye on the target point, not on the line.

2. Ernst Röttger and Dieter Klante, in *Creative Drawing: Point and Line,*[2] suggest a number of patterns that bring doodling out of the conditioned domain of cliché and into the creative realm of visual exploration. Try a few of these patterns; examples are shown in Figure 9-1.

Figure 9.1.

All Drawing Is Memory Drawing

Now direct your visual attention away from the paper image toward objects around you. What is involved in translating a perceptual image of an object in your immediate environment into a reasonable likeness on paper? Drawing from life involves two kinds of memory: (1) the long-term memory necessary to all perception, and (2) the short-term memory that holds a perceptual image in mind while it is being reproduced on paper.

Long-term memory is prone to stereotype. Draw the image of a dog from long-term memory, for example, and it will likely be a cliché image of dogs in general. *When drawing directly from a model, school yourself to avoid the stereotyping effect of long-term memory.* To draw what you see, (1) observe the model for a specific overall relationship or detail, (2) exercise short-term memory to bring that observation to your drawing, (3) refer back and forth between model and drawing frequently.

In the following drawing experience (9-3), reduce the term of your memory to an instant by moving your marker in unison with your eyes. Do not be concerned with drawing a masterpiece; center your attention on simultaneously seeing the object and remembering it on paper. (A drawing exercise that involves increased memory span is included in Chapter 15.)

1. Select a half-dozen objects that you would like to know better, through drawing. Nongeometric objects (such as flowers, vegetables, shoes, your hand) will help to balance the geometric bias in subsequent exercises.

2. Begin by not drawing. Contemplate the object, discovering as much about it as you can. View it all around. Touch it, tap it, smell it.

3. Now relaxedly focus your eyes on the center of the object and loosely place your marker in the center of the drawing-to-be. Eye and marker always working in unison, explore the object visually while simultaneously

building the drawing with scribbles, as shown in Figure 9-2. Don't let your marker get ahead of your eyes; imagine that your marker is a kind of tactile eye that caresses surfaces, probes crevices, and turns corners. Build your drawing freely and vigorously, much as a sculptor would build an object from clay.

4. Check yourself. If you have been expecting to create a minor masterpiece, or even a creditable drawing, you have misinterpreted the intent of this exercise. During the next drawing, be aware only of the moment-by-moment process. Expectations of any sort, including expectations about the end result of your drawing, are ruled by long-term memory; accurate observation involves centering attention in the here-and-now.

Figure 9-2.

Freedom of Gesture

In the previous drawing exercises, did you feel that you drew stiffly? Drawing is essentially recorded gesture: *timid, rigid, and conventional patterns of motor release, or gesture, prevent what you see and feel from flowing freely into what you draw.* Thus the student of Chinese painting and calligraphy is taught that "in no way should the brush be inhibited, neither by a feeble nor a stubborn mind, for freedom is the absolute aim. Freedom of gesture exhilarates. . . . Emancipation of mind and freedom of gesture are in effect identical."[3]

Spontaneity of eye, mind, and hand are clearly evident in the example of calligraphy shown in Figure 9-3. As you do the next exercise (9-4), see if you can begin to loosen up physically, mentally, and emotionally, in emulation of the spirit of this Chinese calligrapher.

Figure 9-3.

1. Relax, especially your arms and hands.

2. Hold your conté stick in an unaccustomed way. Grasp it with all of your fingers, or place it between your second and third fingers. Don't hold it as you usually do when handwriting.

3. On newsprint, express graphically each of the following verbs: Leap / Stumble and Fall / Soar / Struggle / Stretch / Dance / Hit / Ice-Skate / Lift (a heavy object) / Explode. Before drawing, let the feelings that you associate with each action come to the fore; imagine yourself leaping, for example, and feel the leap. Then, without intellectualizing or premeditating, spontaneously draw what you are feeling.

4. Alternatively, listen to music and express your feelings about it on paper. Relax your grip on the marker, relax your mind, and simply let the music flow through you into your drawing.

Right-Brained Drawing

In her excellent book *Drawing on the Right Side of the Brain*,[4] Betty Edwards introduces a number of drawing exercises specifically directed to "fire up" the right hemisphere of the brain. The strategy underlying all of these exercises is "to present the brain with a task which the left hemisphere either can't or won't do." In the exercise from Dr. Edwards's book that follows (9-5), the verbal left brain, accustomed to upright images that it can quickly label, is thwarted in its lazy habits of perception by an upside-down image. Meanwhile, the visually and spatially apt right brain, having little or no need to place the image in a verbal category, happily settles into the pleasant task of reconstructing the image's *structure*.

9•5 UPSIDE-DOWN DRAWING

1. Select an image from a magazine or book. Turn it upside down.

2. Turn on some soft music and relax pleasantly into the timeless state associated with solving puzzles.

3. Beginning at the top of a blank sheet of paper, copy the upside-down image *without once turning it right side up.* As you put the structure of your drawing together, construct it like a jigsaw puzzle, line by line. Focus your attention on lines turning in space, lines joining other lines, and lines outlining shapes that adjoin other shapes to *create a pattern.* Suspend your labeling ability entirely for half an hour while you concentrate on capturing the spatial relationships in a meaningless image.

4. When you are finished, turn your drawing right side up. Surprised how well it came out?

For an advanced version of this exercise, draw an image upside down, so that a person sitting across the table from you would see the image come into being right side up! For a number of other variations on the theme of right-brained drawing, see Betty Edwards's well-written and thoroughly illustrated book.

Draw Things That Interest You

As with all arts, developing the art of seeing takes time and energy. You were born with sight; you must work to develop sight into vigorous seeing. Drawing provides an unequaled catalyst for this, but uninteresting drawing tasks do not. To marshal the additional effort necessary to transform lazy, workaday eyesight into artful seeing, draw things that interest you, as exercise 9-6 suggests.

9•6 INTEREST BOOK

1. On the front page of a bound sketchbook, make a list of objects that interest you. List some of your favorite personal possessions; list objects that interest you professionally. If you are interested in cars, note down several specific models—and so on. At the start, list at least 30 objects; later, expand this list.

2. Every day, sketch several objects selected from your list. Draw each object to the best of your ability. If you are unsatisfied with a drawing, return to draw it again another day—from another viewpoint, with another kind of marker. Most important: draw every day.

References

1. Nicolaides, K. *The Natural Way to Draw.* Houghton Mifflin.
2. Röttger, E., and Klante, D. *Creative Drawing: Point and Line.* Reinhold.
3. Hill, E. *The Language of Drawing.* Prentice-Hall (Spectrum Books).
4. Edwards, B. *Drawing on the Right Side of the Brain.* J. P. Tarcher.

PATTERN-SEEKING

10

Your Pattern-Seeking Nature

You can directly experience your natural tendency to organize visual imagery into coherent patterns by looking at the sea of squares in Figure 10-1. As you look, notice how the squares seem to "group themselves" into swirling patterns. Printed images are obviously inert. *You* are providing

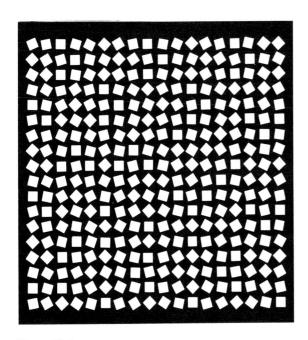

Figure 10-1.

the patterning action that you experience in Figure 10-1.

Pattern-seeking, a natural and important part of every act of visual thought, is the first step of a two-step process: pattern, then analyze. In the *act of seeing*, the first step in this process is to perceive an undetailed, overall pattern; the second step, which proceeds according to personal needs and interests, is to analyze the overall pattern for details. (In lazy and stereotyped vision, as discussed in Chapter 8, seeing proceeds into the second step just far enough to identify the image.) The two-step, pattern-then-analyze process is readily observed in the *act of drawing:* first, the visualizer "roughs in" the drawing's overall relationships; second, he or she develops the drawing's details, again seeking to bring into focus what is perceived as important. Again, in the *act of imagining,* we find the same two-step process: inner images that are initially abstract and stereotyped, for example, can be clarified by attending to the details in the image with all of your senses (see Chapter 15, Visual Recall).

Center your attention now on seeing the patterns, or in psychological terminology, on the "gestalten" of visual imagery. In the next chapter, you will take the natural next step, seeing analytically.

The Gestalt

"Gestalt" is a German word that has no exact equivalent in English. Form, shape, configuration, and pattern come close; organizational essence perhaps comes closer. Toward the end of the nineteenth century, a group of Austrian and German psychologists began to perform research and formulate theories about the role of pattern-seeking in human behavior. Gestalt psychology, which evolved from these studies, has been especially productive in the field of visual perception.

Gestalt psychologists hold that perception inherently acts as an active force, comparable to a magnetic field, that draws sensory imagery together into holistic patterns, or "gestalten." According to this view, *every perceptual image consists of more than the sum of its parts; it also possesses a "gestalt," a patterning force that holds the parts together.*

The example of a six-note melody is sometimes used to clarify this notion. The six notes of the melody are its parts. With considerable freedom, you can change these melodic parts without changing the melody itself. You can, for example, change key, move up an octave, modify the rhythmic phrasing from waltz to bossa nova. Through all of these changes, the melody remains the same. The melody is the forceful "seventh part" that holds the six-note musical phrase together. The melody is the phrase's gestalt.

Fingerpainting provides an excellent medium for experiencing the gestalt, or "melody," of visual imagery: it discourages overconcern with detail, encourages rapid image formation, and deeply involves the visualizer in creative patterning. In the next exercise (10-1), consider not only the gestalt of the image in the fingerpaint but also the larger gestalt, or harmonious whole, that relates you to the model and to the image that you are forming on the paper. Creative people frequently report that they enter into, and even become, the object of their creation. As you work with the fingerpaint, feel with the action of the model that you are viewing, and bring that feeling directly into the fingerpaint image.

10•1 FINGERPAINT PATTERNS

A live model is recommended for this exercise, but in the absence of a model, magazine photographs will do.

1. Protect table and clothes with newspapers (which can be pinned to clothes).

2. Pour a few tablespoons of dark commercial fingerpaint (or dry tempera mixed with liquid starch) onto a large sheet of glazed shelf paper or butcher paper.

3. Explore the delicious messiness of the medium for a few minutes. Use both hands, fingernails, a comb. Create spatial fantasies. Interpret music. Be a child again.

4. Ask the model to take a series of action poses (reach, bend over, kick), changing every 30 seconds.

5. Look at the model and feel his or her action in your own body. Feel what you see; then capture that feeling in fingerpaint. If the model is reaching, feel reaching as you draw. "Draw not what the object looks like, not even what it is, but *what it is doing.* . . . Try to feel the entire thing as a unit—a unit of energy, a unit of movement."[1]

6. Don't outline the model. Don't fuss with details. Be concerned only with basic relationships. Move your entire hand through the center of the image, quickly and rhythmically capturing the gestalt of the pose—nothing more.

7. To preserve an image, place a sheet of newsprint over the fingerpaint, rub it smoothly, and remove a "mono-print."

Grouping

According to Gestalt theory, perception obeys an innate urge toward simplification by assembling complex stimuli into simpler groups. Three grouping effects, shown in Figure 10-2, are grouping by proximity, similarity, and line of direction. On the left of the figure, you undoubtedly see two groups of patches. "Why not merely six patches? Or two other groups? Or three groups of two members each? When looking casually at this pattern everyone beholds the two groups of three patches each."[2] The six patches are grouped into two clusters by *proximity*. In the center, the white and black discs are grouped into a line and a triangle by *similarity*. On the right, the random shapes are grouped into a serpentine by a *line of direction*.

Grouping occurs involuntarily. In Figure 10-1, we cannot help but see the squares grouped along lines of direction. Similarly, we do not have to decide to perceive the many leaves of a tree as a single mass of foliage, or the thousand windows of a skyscraper as a single fenestration pattern. Our nervous system automatically groups these visual complexities for us.

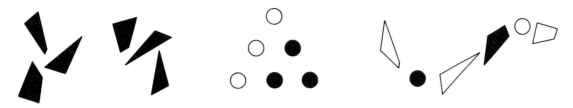

Figure 10-2.

Grouping can also be imposed voluntarily. Painters consciously group visual elements to obtain unity in their paintings. Scientists group, or classify, their observations. Students group their notes into outline form. "The binding fact of mental life in child and adult alike," observes Jerome Bruner, "is that there is a limited capacity for processing information—our span, as it is called, can comprise six or seven unrelated items simultaneously. Go beyond that and there is overload, confusion, forgetting."[3] A primary method for organizing information together into an attentive whole is grouping.

In the next exercise (10-2), voluntarily use the grouping of principles of proximity, similarity, and line of direction to simplify the image and to find its underlying, unifying pattern.

10·2 GROUPING

1. Overlay a photograph or drawing with a piece of tracing paper. Without regard for detail, boldly use a felt-tip marker to represent the five to eight groupings in the image that constitute its unifying gestalt. Parts of a human figure, for example, are grouped by lines of direction. Several pictures on a wall are grouped by proximity. Other forms are grouped by similar color or shape. Block in these groups, while also seeking the overall relationship that unifies the groups. Try several versions. Grouping is creative; there's more than one way.

2. Alternatively, find basic groupings in the compositions of paintings, or in the patterns of actual landscapes, cityscapes, room interiors, or single objects. Also create alternative grouping relationships for the same image.

3. Working quickly, do many of these drawings. Grouping of this sort is an act of invention that is extremely important to visual thinking.

Tachistoscopic Seeing

Another way to exercise your capacity to see visual wholes is provided by the tachistoscope, an instrument for flashing images on a projection screen. Tachistoscopic images are used by the military to teach personnel to recognize instantly the silhouettes of various kinds of aircraft and ships. Aldous Huxley, in an essay in which he argues for educational innovation, points out that "old habits of distorted and conceptualized seeing are bypassed by the flashing magic lantern. The student recaptures his visual innocence; for a hundredth of a second he sees only the datum, not his self-dimmed and verbalized notion of what the datum should be."[4] To put the point another way, flashed images bypass the verbal left brain and call forth the powers of the right brain to perceive and conceive spatial relationships.

You need not purchase an expensive commercial "T-scope" to experiment with tachistoscopic training. You can simulate a T-scope shutter by blocking the light beam from a slide projector by hand, exposing the image with a flick of the wrist. Or, in a darkened room, you can flash a strobe light on a live model. Without equipment entirely, you can look toward an object or scene with your eyes closed, and "flash" the image by blinking your eyes. Whatever your means, flash the image quickly, in the range between one-tenth and one-hundredth of a second.

Figure 10-3.

A competent amateur photographer can prepare slide material for T-scope exercises by following these guidelines:

1. Photograph highly contrasted "figure-ground relationships."

2. Avoid subject matter or viewpoints that are ambiguous.

3. Avoid strong shadows that obscure form.

In tachistoscopic viewing we record visual experiences "nearly instantaneously and without participation of conscious processes"; later we may represent them in such behavioral responses as "doodling."[5] One form of doodling well-known to artists and designers is the "thumbnail sketch." *Quickly executed "thumbnails," by virtue of their smallness, are well suited to capture the visual essence of an image.* Smallness gives the illusion of distance in which details merge into overall relationships; drawing in a small area facilitates eye-hand coordination; image unity is easier to capture in a small drawing than in a large one. The "thumbnail" in Figure 10-3 was drawn by architect Eric Mendelsohn in preparation for the design of a building.

10•3 T-SCOPE THUMBNAILS*

Draw thumbnail sketches from flashed images. Flash the image, draw, and then compare your drawing with the image. Build up the image loosely and quickly. Capture basic groupings. Develop the entire image at once, without fussing over detail, into a unified whole.

Meaningful Patterns

The grouping principles described so far are concerned with unifying visual patterns without regard to their representational meaning. The tendency of perception to seek meaningful patterns is demonstrated by Figure 10-4.

*This use of the T-scope to educate seeing was pioneered by Hoyt Sherman of Ohio State University.

Figure 10-4.

This classic illustration has two meanings: it can be seen as a vase, or alternatively, as two human profiles. Notice, as you view this design, that one or the other of these meanings come to the fore. When you see the white vase, the black shape has no meaning except as background. When the black shape advances in space to become two profiles, the previous vase retreats into meaninglessness. You cannot perceive both meanings at the same instant. *Your pattern-seeking nature, in its quest for meaning, craves unity in a single figure-ground relationship.*

Vision also seeks spatial meaning. The configuration in Figure 10-5, called "Thiery's figure," has two spatial meanings: you can perceive that you are looking either up or down at the horizontal uncheckered planes. And, as in the previous figure, you can perceive only one of these meanings at a time.

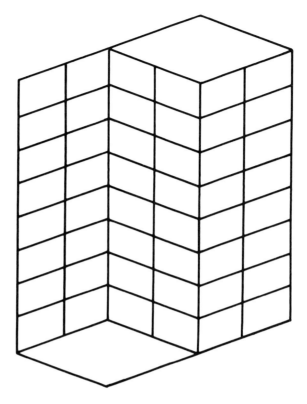

Figure 10-5.

The amusing drawing in Figure 10-6 presents an ambiguous figure-ground relationship that frustrates your natural urge to seek a meaningful pattern.

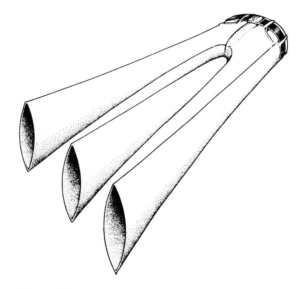

Figure 10-6.
Levi Strauss & Co./Honig-Cooper & Harrington.

So powerful is your perceptual tendency to perceive meaningful patterns that you will fill in missing parts. This grouping effect, known as closure, was discussed in Chapter 3 and illustrated in Figure 3-1. Artists frequently use closure to provoke the imagination of the viewer. The drawing reproduced in Figure 10-7, by its artful incompleteness, makes the viewer complete the image by an act of imagination.

Professor Charles Rusch has developed the following exercise (10-4) to dramatize how we visually process patterns for meaning. An excellent supplement to the previous exercise, it also requires a projector.

10•4 FOCUS!

1. Project a slide on the screen as far out of focus as the focus control permits.

2. Slowly and gradually bring the slide into focus.

3. Examine the process by which you find meaning in the image. How does the unfocused, meaningless pattern make you feel? As the pattern comes into focus, do you perceive meaning prematurely and find later that you are mistaken? What is the effect of such preconception on subsequent seeing?

With experimentation, you will find that certain slides can be especially tantalizing: for example, the initial unfocused image that strongly suggests a misleading meaning, or the unfocused pattern that dramatically regroups in the final stage of focusing, or the unfocused pattern that is strikingly beautiful until it comes into focus as a decidedly ugly object.

Projection

Projection is a perceptual phenomenon similar to closure. Faced with an incomplete and amorphous pattern such as the inkblot in Figure 10-8, we tend to rummage about our imagination until we find a meaningful image that we can project onto that pattern. Projections are often fanciful: one person viewed an inkblot and saw "a lanky boy and a jester watching the antics of an inebriated abbot." Projections are also highly related to personal interest: a woman discovered a "bonnet with feathers" and a minister "Nebuchadnezzar's fiery furnace" in the same inkblot.[6] The

Figure 10-7.

well-known Rorschach inkblot test, by stimulating projection, enables psychotherapists to discover the subconscious interests of their patients.

Figure 10-8.

Projection has played an important role in the development of art. The eminent art historian H. W. Janson observes that in some early cave pictures

the shape of the animal seems to have been suggested by the natural formation of the rock. . . . We all know how our imagination sometimes makes us see all sorts of images in chance formation such as clouds or blots. A Stone Age hunter, his mind filled with thoughts of the big game on which he depended for survival, would have been even more likely to recognize such animals as he stared at the rock surfaces of his cave. . . . Perhaps at first he merely reinforced the outlines of such images with a charred stick from the fire, so that others, too, could see what he found.[7]

And Leonardo da Vinci made this note about projection:

I cannot forbear to mention . . . a new device for study which, although it may seem trivial and almost ludicrous, is nevertheless extremely useful in arousing the mind to various inventions. And this is, when

you look at a wall spotted with stains . . . you may discover a resemblance to various landscapes, beautified with mountains, rivers, rocks, trees . . . or again you may see battles and figures in action, or strange faces and costumes, and an endless variety of objects which you could reduce to complete and well-drawn forms. And these appear on such walls confusedly, like the sound of bells in whose jangle you may find any name or word you choose to imagine.[8]

In the absence of "a wall spotted with stains," create your own foil for projection (10-5).

10•5 DA VINCI'S DEVICE

A quick way to obtain an amorphous pattern "useful in arousing the mind to various inventions" is to scribble lines on a piece of paper.

1. Close your eyes (so that you will not be tempted to influence the scribble) and cover a sheet of newsprint with a random network of lines. Use a light gray marker, such as a gray felt-tip pen.

2. Open your eyes and discover resemblances in the scribble. If you find that projected images do not come easily to you, seek specific meanings (faces, birds, animals). Once you've had this practice, open yourself up to whatever meaningful patterns come forth unbeckoned.

3. Reinforce and develop the meaningful patterns with a black nylon-tip marker or with color.

Pattern-Seeking and Problem-Solving

The patterning principles of grouping, projection, and closure are especially important to visual problem-solving. As mentioned earlier, pattern-seeking is the first step of all perception. *The pattern, or gestalt, that you perceive in a problem strongly influences the way you attempt to solve that problem.* Stereotyped thinkers work with the first pattern that they see, almost invariably a conventional one. Creative thinkers recenter their perception of the same problem by regrouping it into a variety of patterns. Edward de Bono, in *New Think*, creates the term "vertical thinking" for thinking that begins with a single perceptual pattern and proceeds immediately to delve deeply into that pattern for a solution. Thinking that generates alternative ways of seeing the pattern of a problem before seeking a solution he calls "lateral thinking."[9]

De Bono diagrammatically illustrates the regrouping principle underlying lateral thinking by the patterns reproduced in Figure 10-9. Assume that the problem in Figure 10-9 is to describe the shape numbered 1 with the minimum number of words and maximum clarity, as though you were writing a telegram. Unlike the vertical thinker, the lateral thinker would perceive several alternative groupings (patterns 2 through 6) before choosing the easiest one to describe verbally. According to de Bono, *recentering by repatterning opens possibilities to creative and unusual solutions* that are closed to the thinker who enters into problem-solving by relying on the first pattern that comes to mind.

The pattern-seeking effect of projection, by bringing inner thoughts into the act of seeing, can also aid problem-solving very directly. A sculptor, for example, may look at the grain pattern in a piece of wood and suddenly "see" the sculptural form that he or she will then attempt to realize. Projection commonly contributes answers to externalized thinking in every field, from technological invention to choreography (see Chapter 7).

Finally, problem-solving is often experienced as seeking closure in an incomplete pattern, as illustrated by the following puzzle (10-6).

10•6 INCOMPLETE FRET

The fret, or latticework, in Figure 10-10 is incomplete. Can you capture the spirit of the pattern and turn the corner that the designer left unfinished? Place a piece of tracing paper over the pattern and fill in the missing black areas; "interlace the bands by weaving them alternatively under and over each other."[10]

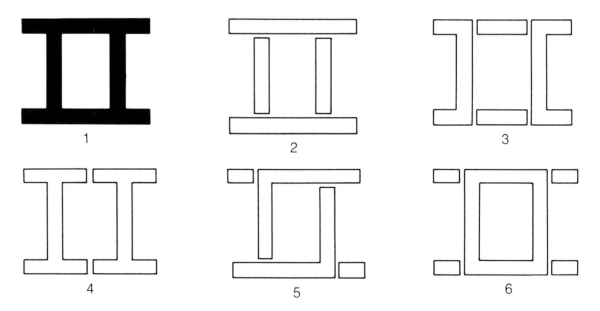

Figure 10-9.

Figure 1-6 of *New Think* by Edward de Bono, © 1967, 1968 by Edward de Bono, Basic Books, Inc., Publishers, New York. Reprinted with permission of the publisher and Edward de Bono, author of *The Mechanism of Mind* (Simon & Schuster), Lateral Thinking for Management (American Management Association), and *Beyond Yes and No* (Simon & Schuster).

From *The Book of Modern Puzzles* by Gerald L. Kaufman, Dover Publications, Inc., New York, 1954. Reprinted through permission of the publisher.

Figure 10-10.

Pattern, Then Analyze

Our ability to find meaningful wholes in visual imagery is complemented by our capacity to analyze, to divide wholes into parts. To see fully and creatively requires both patterning and analytical abilities. If you normally tend to analyze imagery, to dwell on and develop details, then discipline yourself to see the gestalt that brings parts into a unified whole. Draw from flashed images; sketch "thumbnails"; school yourself to see groupings *and regroupings* in visual patterns. Conversely, if "big brush" imagery and thinking come easily, develop the analytical concern for visual detail that likely does not; the next chapter concentrates on analytical seeing.

In either case, remember that the phrase "pattern, then analyze" describes the natural sequence of all visual-thinking processes. No amount of careful, detailed analysis can remedy an overall visual pattern that has been inaccurately seen or conceived.

References

1. Nicolaides, K. *The Natural Way to Draw.* Houghton Mifflin.
2. Kohler, W. *Gestalt Psychology.* New American Library (Mentor Books).
3. Bruner, J. *On Knowing: Essays for the Left Hand.* The Belknap Press of Harvard University Press.
4. Huxley, A. *Tomorrow, Tomorrow and Tomorrow.* Harper & Row.
5. Kubie, L. *Neurotic Distortion of the Creative Process.* Farrar, Straus & Giroux (Noonday Press).
6. Bartlett, F. *Remembering.* Cambridge University Press.
7. Janson, H. *History of Art.* Prentice-Hall.
8. Da Vinci, L. *The Notebooks of Leonardo da Vinci* (edited by Taylor). New American Library (Mentor Books).
9. De Bono, E. *New Think.* Basic Books.
10. Kaufman, G. *The Book of Modern Puzzles.* Dover.

Recommended Reading

These well-written and profusely illustrated books are concerned with the organizational principles (or "grammar") that make visual expression possible:

Anderson, D. *Elements of Design.* Holt, Rinehart & Winston.
Arnheim, R. *Art and Visual Perception.* University of California Press.
Collier, G. *Form, Space, and Vision.* Prentice-Hall.
Dondis, D. *A Primer of Visual Literacy.* The MIT Press.
Gombrich, E. *Art and Illusion.* Pantheon Books.
Kepes, G. *Language of Vision.* Paul Theobald.

ANALYTICAL SEEING

11

Seeing the Details

"Ordinarily we do not make full use of our faculty for seeing," writes Edward Hill.[1] Instead, we usually make habitual, computer-like responses to visual information, automatically selecting out what is immediately relevant to our needs, and little else. At this utilitarian level, vision is impoverished: we look, we act, but we do not see. Or more accurately, we lazily see only stereotyped visual concepts. In this chapter, *you will experience going beyond routine pattern-seeking to explore the details that make each visual image unique.*

Seeing fully takes time. Observes Ross Parmenter, in *The Awakened Eye,*

> With large, detailed gestalten we need lots of time: time to absorb the gestalt itself so that we can take it for granted and so loosen its grip; time to segregate it into sections that we can examine individually; and time to reformulate it so we can get a new total idea of it, with awareness of the working of its parts, as well as of the integration of the whole?

Take wine tasting, for example. If you are an experienced wine taster, you enjoy far more than a "drink." First, you probably

assimilate an overall sensory impression. You then carefully consider the wine's individual qualities (its bouquet, for example, or clarity), comparing these with your memory of like qualities in other wines. After discriminating the character of each sensory detail, you fuse this new knowledge into a new gestalt, a more complete overall impression than was possible at first taste. Visual experience that goes beyond stereotypes involves a similar kind of energetic and joyful connoisseurship.

Seven Million Colors

According to psychologist Jerome Bruner, the human eye can discriminate seven million hues and shades of color. Look around you right now; does any object in your room reflect a *single* color? In my view now is a "yellow" chair that, in fact, reflects a subtle gradation of yellows, a gradation that will change continually during the day and shift markedly as day turns into night. Ordinarily, people do not discriminate such differences in color; they simply see the chair as "yellow." As Bruner observes,

> Were we to utilize fully our capacity for registering the differences in things and to respond to each event as unique, we would soon be overwhelmed by the complexity of the environment. . . . The resolution of this seeming paradox—the existence of discrimination capacities which, if fully used, would make us slaves to the particular—is achieved by man's capacity to categorize. In place of a color lexicon of seven million items, people in our society get along with a dozen or so commonly used names. It suffices to note that the book on the desk has a "blue" cover.[3]

It may suffice for the verbal thinker to reduce visual experience in this way, but not for the visual thinker.

According to Bruner, *our experience of color is diminished and simplified in two ways: by color constancy and color labeling.* By color constancy, we see a white house immutably and all-over white, whether we see it in bright sunlight, in the rosy light of sunset, in moonlight, or with one wall brightly lighted and another in deep shadow. By color labeling, we categorize and label colors with words: hundreds of shades of white are merely "white." Visual thinkers lift the averaging filter of color constancy from their eyes to see the subtle effects that lighting plays on color. And they lift the categorizing filter of color labeling so that they can see the richness of color that verbal labeling conceals.

One of the best ways to overcome the stereotyping effect of color constancy and color labeling, and to see color afresh, is through color matching. The artist who mixes colors on a palette to match the colors in the sky knows that the sky is not merely "blue" but a subtle gradation of colors. The following exercise (11-1), suggested by Ross Parmenter, uses color matching to heighten sensitivity to the luxurious variety of color that eludes lazy, workaday seeing.

11•1 PAINT CHIP HUNT[2]

1. Obtain several dozen paint chips from your local paint store. (They're samples and usually free.) Ignore the labels on the back.

2. Now the hunt: match, as exactly as you can, each paint chip with an object in your environment (such as clothing, plants, furniture, building materials, supermarket products, or advertisements). Matching the more exotic colors may be a challenge, but don't get discouraged. Be aware of the analytical nature of color matching (this red is slightly more orange than that one; this blue is lighter than any of those).

The Tactile, Kinesthetic Eye

Seeing only the gestalten of visual imagery, without also seeing the diverse detail embedded in these larger patterns, is much like gulping down food without savoring it. Eating by large bites, we miss the nuances of flavor and texture available to the gourmet, who assimilates a meal slowly, analytically, and with all senses engaged. The visual sense, partially because we use it so much, is prone to con-

suming experience only by large bites. But we can use the nonvisual senses (which are less capable of perceiving large gestalten) to help us direct our perception toward discriminating particular features. Nature does not separate seeing from the other senses; only words do. *Seeing is polysensory, combining the visual, tactile, and kinesthetic senses.*

Of the tactile eye, Rudolf Arnheim writes, "In looking at an object we reach out for it. With an invisible finger we move through the space around us, go out to the distant places where things are found, touch them, catch them, scan their surfaces, trace their borders, explore their texture."[4] A few people have a natural predilection for perceiving primarily by their sense of touch; psychologists call them "haptics." The haptic person has adequate vision, but defers vision to touching. Many more people are conditioned to have the opposite tendency. Unlike the haptic, they experience the world as if it were everywhere posted with signs warning "Do not touch."

Of the kinesthetic eye, William James said, "The muscular sense has much to do with defining the order of position of things seen, felt, or heard. We look at a point; another point upon the retina's margin catches our attention, and in an instant we turn our attention upon it." Such visual scanning, involving the muscles of the eye, is similar to exploring an object by hand, the way a blind person does. In visual as well as tactile-kinesthetic perception, "the shape is abstracted from the object by virtue of the actions which the subject performs on it, such as following its contour step by step."[5]

In the next exercise (11-2), begin by not using your eyes. Explore objects as if you were a haptic or blind person.

11•2 FEELIES

1. Have someone put several objects that are unknown to you into separate, good-sized paper bags.

2. Without looking, reach into a bag and explore

the object inside, tactually and kinesthetically. Take time to perceive the form thoroughly: its edges, corners, roundnesses, concavities, roughnesses, smoothnesses, temperature, weight—its spatial configuration.

3. Still without looking, transform your tactile-kinesthetic perception of the object into visual form by drawing it. As you draw, feel the object as much as you like, but don't peek!

4. Unbag the object and compare it with your drawing. Correct the drawing with a marker of a different color.

5. Repeat with another object.

The previous exercise dramatizes the point that vision does not occur alone. Vision is actually polysensory: you see not just patches of color but objects that are hard or soft, warm or cold, rough or smooth, light or heavy. Your eyes sense these qualities because your visual, tactile, and kinesthetic senses are fused. Further, by bringing the tactile and kinesthetic senses into play, the analytical nature of seeing comes into balance with the equally natural tendency to seek overall and undetailed visual patterns.

Kimon Nicolaides observed in *The Natural Way to Draw* that "learning to draw is really a matter of learning to see—to see correctly—and that means a good deal more than merely looking with the eye. The sort of 'seeing' that I mean is an observation that utilizes as many of the five senses as can reach through the eye at one time."[6] In the following drawing exercise (11-3), devised by Nicolaides, merge vision with touch and kinesthesia.

11·3 CONTOUR DRAWING

Place one of the objects used in the previous exercise on the table before you. With black nylon marker on newsprint, make a full-size drawing of the object, as follows:

1. "Focus your eyes on some point—any point

will do—along the contour of the model. (The contour approximates what is usually spoken of as the outline or edge.) Place the point of the marker on the paper. Imagine that your marker point is touching the model instead of the paper. Without taking your eyes off the model, wait until you are convinced that the marker is touching that point on the model upon which your eyes are fastened.

2. "Then move your eye slowly along the contour of the model and move the marker slowly along the paper. As you do this, keep the conviction that the marker point is actually touching the contour. Be guided more by the sense of touch than by sight. *This means that you must draw without looking at the paper,* continuously looking at the model.

3. "Exactly coordinate the marker with the eye. The eye may be tempted to move faster than your marker, but do not let it get ahead. Consider only the point that you are working on at the moment with no regard for any other part of the figure.

4. "Not all of the contours lie along the outer edge of the figure. . . . As far as the time . . . permits, draw these 'inside contours' exactly as you draw the outside ones. . . . *Develop the absolute conviction that you are touching the model.* Draw . . . slowly, searching, sensitively. Take your time."

5. Repeat the exercise, making a contour drawing of your hand or your shoe. Remember: the desired result of this exercise is an experience in analytical seeing, not a masterpiece.

Seeing and Knowing

As discussed in Chapter 8, the knowledgeable observer sees things that a less knowledgeable companion literally cannot. Thus Norwood Hanson notes in *Patterns of Discovery*, "The infant and the layman can see: they are not blind. But they cannot see what the physicist sees; they are blind to what he sees."[7]

Far more than the full participation of the senses is involved in analytical seeing. The intellect, with its vast store of knowledge, is also involved, and intimately so. And because so much knowledge is stored in relation to language, *words can powerfully catalyze seeing*. Scientific observers are especially alert to the way careful verbal description brings knowledge to play and thereby makes seeing more accurate.

Not any sort of language will do, however. Cliché labeling leads only to cliché seeing. As Parmenter puts it, "Don't chip away at things to make them fit words, but instead conscientiously use words to try to make them fit things." Such a search for precise verbal description does three things: (1) it enhances visual memory by relating visual imagery to existing verbal knowledge, (2) it disciplines seeing by joining verbal and visual searching together, and (3) it educates ambidextrous thinking (see Chapter 4).

11·4 VERBAL SEEING

In addition to drawing, describe what you see with words. Talk to yourself or take notes, being careful to be precise. Ask yourself "How would I describe this to someone who hasn't seen it?"

If possible, ask one member of a group to describe a simple object with words, and then ask the remaining members of the group to draw the hidden object from the verbal description only.

The person whose use of words is most knowledgeable, imaginative, and complete will, of course, be best able to use words to enliven seeing. To believe that seeing is merely a matter of turning on the senses, and not also turning on the intellect and the unique human capacity to elicit imagery with words, is to miss a crucial point. Seeing is encountering reality with all of your being. To encounter reality deeply, you cannot leave part of yourself behind. All of your senses, your emotions, your intellect, your language-making ability—each contributes to seeing fully.

Analyze, Then Repattern

In the last chapter you were advised to "pattern, then analyze." In the pattern-seeking mode you are better able to choose what is worth observing. Also, when focused on patterns, the eye is more apt to catch changes, differences, and the unexpected. The Samurai swordsman, to whom the unexpected could be a matter of sudden death, was carefully trained to keep his senses open to the largest possible pattern. With his attention focused broadly, he was more apt to detect a surprise attack than when his attention was focused analytically on a single object.

If pattern-seeking should precede analytical seeing, so should visual analysis be followed by a reformulation of the overall pattern. Indeed, this is a fundamental way to take visual gestalten out of the musty realm of stereotype. After you have analytically seen a hundred different shades of blue, the single category "blue" is never quite the same again. Pattern-seeking and analytical seeing are the basic two phases of seeing (and of visual thinking generally). By cycling back and forth between the two—patterning, then analyzing, then repatterning—you more fully exercise and utilize your visual and mental capabilities. Such an interaction is involved in the perception of proportional relationships, as you will experience in the next chapter.

References

1. Hill, E. *The Language of Drawing*. Prentice-Hall (Spectrum Books).
2. Parmenter, R. *The Awakened Eye*. Wesleyan University Press.
3. Bruner, J., Goodnow, J., and Austin, G. *A Study of Thinking*. Wiley.
4. Arnheim, R. *Visual Thinking*. University of California Press.
5. James, W. *Psychology: The Briefer Course*. Harper & Row (Harper Torchbooks).
6. Nicolaides, K. *The Natural Way to Draw*. Houghton Mifflin.
7. Hanson, N. *Patterns of Discovery*. Cambridge University Press.

PROPORTION

12

Seeing Things in Proportion

Sensitivity to proportional relationships pervades everything we see, think, or do. "Making a mountain out of a molehill" is a failure to perceive relationships as they are. From the perception of the problem to the evaluation of the final solution, seeing things in proportion, involving both pattern-seeking and analytical modes of perception, is vital to effective visual thinking.

People vary greatly, however, in their ability to perceive proportional relationships. Psychologist I. Macfarlane Smith found, for example, that some test subjects, asked to copy a simple shape, not only produced a drawing that was grossly out of proportion, but also appeared to be unable to perceive the proportional discrepancies between their copy and the original.[1] Jay Doblin, reflecting on his experience teaching drawing to design students, observes: "The most important and difficult mental control is the judgment of proportion. The mental processes that control accurate proportion on paper are the same as those that permit the designer to judge good proportion in objects."[2]

In the following exercise (12-1), test your current ability to perceive accurately the proportion of simple shapes.

12·1 PROPORTION OF SIMPLE SHAPES

1. With black marker on newsprint, draw several simple geometric and free-form shapes of your invention (about 3" in size).

2. Place a sheet of tracing paper over one of the shapes and trace approximately one-quarter of its outline. Remove the tracing paper.

3. Place the partial tracing alongside the newsprint and copy* the remaining outline of the shape.

4. Superimpose the two outlines. Trace the original shape with a blue marker to clarify errors in perceiving proportion.

5. Repeat, or go on to the next shape. If you experience difficulty with this exercise, repeat it frequently. The ability to record proportion improves rapidly with the accurate reinforcement that this exercise provides.

Superimposition

Judging proportion requires the assessment of relationships between parts: proportion is a matter of ratios. A primary way to judge proportional ratios is superimposition. Asked to judge the proportion of a rectangle, for example, you would most likely superimpose one side on another, arriving at a ratio such as 1 to 4.

A classic superimposition device in figure drawing is the human head, as shown in Figure 12-1. Indeed, *any simple and distinct part of an image can be visually superimposed to measure proportional relationships of the whole.* Wheels can be superim-

*Making copying taboo, as it is in some contemporary art classes, is as wrong as making copying an obsession, as was the academic practice not too long ago. Copying does not stifle creativity when it is used selectively to educate vision. Creative artists throughout history have educated themselves by copying.

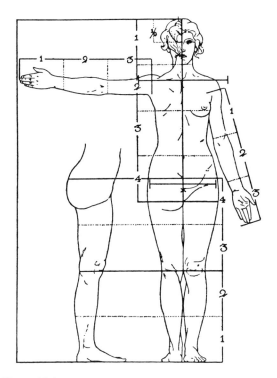

Figure 12-1.

Figure 12-2.

posed to measure the proportions of an automobile; a bridge's tower can be superimposed to measure its span. A fundamental scheme for judging proportion by eye involves superimposing an internal part on the whole. An external measurement device can also be used. Thus artists commonly superimpose a pencil held at arm's length to gauge proportions. They may also use a square viewer—a small square hole cut in stiff paper.

When a shape is complex, you can also help the eye judge proportion by superimposing a grouping scheme upon the whole. By grouping, you reduce the whole to simpler parts and more easily comprehended ratios, as shown in Figure 12-1.

Another analytical scheme for assessing proportional relationships is the superimposed square grid suggested in Figure 12-3. A square grid drawn on tracing paper can actually be superimposed over a shape, or a square grid can be fixed in visual memory by exercises such as the next two (12-2, 12-3). Once committed to memory, square grids can be projected mentally to assess proportion. Indeed, they automatically will be.

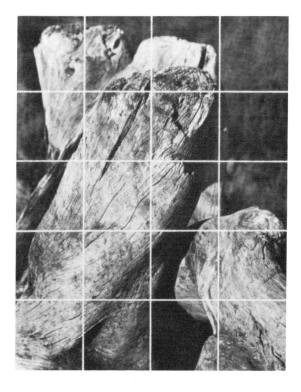

Figure 12-3.

12•2 MULTIPLE SQUARES

1. Cut a sheet of newsprint, along the long direction, into a number of 2'' ribbons.

2. Estimating proportion by eye, snip the ribbons into rectangles having the following dimensions: 2 × 2, 2 × 16, 2 × 4, 2 × 14, 2 × 6, 2 × 12, 2 × 8, 2 × 10. In your imagination, superimpose multiple squares upon the ribbon before you cut it.

3. With a ruler, measure the length of each rectangle. Mark proportion error on each (for example, ½'' short). Repeat until error is small.

4. On a second sheet of newsprint, draw a similar variety of rectangles (freehand, by eye). Check your accuracy with a strip of paper marked with the short dimension of the rectangle.

12•3 SUPERIMPOSED GRID

1. On tracing paper, using a ruler, triangle, and blue marker, draw two 5'' squares subdivided into a grid of 1'' squares. The two grids should be separated 2'' apart.

2. Place a second sheet of tracing paper over the grids.

3. With a black marker, draw a simple shape of your invention within the left-hand 5'' grid (right-hand grid if you are left-handed). This shape is your "model."

4. Over the adjacent grid, copy the model. Use both grids to guide you.

5. On a separate sheet of newsprint, copy the model once again. This time superimpose the grid only in your imagination.

6. Check the accuracy of the last drawing by slipping it under the grid and comparing it with the model.

7. Repeat. Progressively increase the complexity of the initial shape. With practice, your ability to do this exercise will improve.

However slavish the previous exercises may seem, they do have the virtue of accurate self-reinforcement. Unless you know that you are seeing proportion inaccurately and practice diligently to correct your errors, proportional misperception will plague every aspect of your visual thinking.

12•4 EVERYDAY PROPORTION

Practice daily your ability to see things in proportion. In your imagination, superimpose (1) a grid, (2) a part of the whole, or (3) a grouping scheme upon your coffee cup, house, bed, breakfast table, cat, or whatever. Develop the habit of seeing everyday objects in proportion.

With practice, superimposition devices become built into the imagination, and accurate perception of proportion becomes less analytical and more allied to overall pattern-seeking (see exercise 12-4). Le Corbusier, a master of architectural proportion, wrote: "For fifteen years, putting mathematics and geometry into what I do has been a regular and natural part of my training, a quite simple and spontaneous part. Eye and hand have become expert; consequently I instinctively put things in proportion, but I do not stop watching over the job of getting them exactly right."[3]

Le Corbusier's desire to get proportions "exactly right" suggests that he took great care with the subtleties of proportional relationships. A slight adjustment here, an infinitesimal shift there: subtle changes can have a surprisingly powerful effect on the visual identity and aesthetic quality of the image, as you will see in the next exercise (12-5).

12•5 SUBTLE CHANGES

1. Choose an existing image: a face in a magazine, perhaps something that you are designing.

2. Draw or trace a frontal (none-foreshortened) view of the image.

3. Side by side on tracing paper, draw a series of images that register subtle changes in the proportions of the original image. For example, for the frontal view of a face, slightly enlarge the chin and trace the remainder of the face as is. Alongside, in a second tracing, make the eyes slightly smaller—or closer together. Making only one basic change per image, shorten the forehead; make the lips fuller; lower and thicken the eyebrows; lengthen the ears; and so on.

4. Compare the images and sensitize yourself to the potency of subtle changes of proportion.

Figure 12-4.
Oliver Wendell Holmes. Detail from a drawing by David Levine in *Life Magazine*, October, 1971. Reprinted with permission of David Levine.

Distortion

Another way to sensitize the eye to proportion is to make deliberate distortions, as a caricaturist does to dramatize identity. By exaggerating the proportion of a nose or a chin, the caricaturist seizes upon the gestalt of a face and somehow intensifies it. The caricature in Figure 12-4 induces an emotional response in the viewer partially because feelings are very readily aroused by a departure from what is considered "normal" or "visually correct." Artists also consciously and unconsciously use distortion to evoke emotion and to create aesthetic tension between the distorted image and the "normal" image of the subject that the viewer carries in memory.

As a caricaturist uses distortion to bring you to see a familiar face in a new way, *you can use the technique of changing proportional relationships in general to re-center the way you see things.* Experimental play with proportional relationships, of the sort suggested in the next exercise (12-6), is extremely important to flexible visual thinking.

12•6 CARICATURE

Draw a caricature of yourself or a friend. Boldly—very boldly—exaggerate the subject's distinguishing features. Also use distortion to capture your feelings about the person, or to evoke how you believe the person is feeling.

Functional and Aesthetic Proportion

Now let us apply the concept of proportion to considerations of function and aesthetics. Functional proportion, unlike aesthetic proportion, is usually reducible to numbers. The structural function of a beam, for example, is related to the numerical ratio of its dimensions; the useful function of a chair is related to the measured height of the seat above the floor. Change these measurements and you will likely drastically modify function or even alter functional identity. Size the beam's cross-section smaller and you will cause it to fail; lengthen the chair's legs and you have a bar stool. Few designers, however, determine functional proportion entirely by numbers. Experienced designers have a feel for functional proportion. Especially at the crucial first stages of design, they size functional elements qualitatively, not by numbers but by eye.

Aesthetic proportion, the harmonious visual relationship of parts to the whole, is essentially qualitative, despite many attempts to quantify it. The ancient Greeks, for example, used geometrically derived golden-section rectangles to design the proportions of the Parthenon; medieval architects determined aesthetic proportions with modular schemes based on equilateral triangles and squares, which were thought to possess mystical meanings; the unit of proportion of many Renaissance buildings is one-half the diameter of the columns.

Such attempts to quantify aesthetic proportion must contend with the perceptual field theory of Gestalt psychology. When apparent size (or any other visual quality such as color) is profoundly influenced by the context in which it is situated, as is illustrated by the optical illusion in Figure 12-5 (the center circles are the same size), quantitative aesthetic ratios are clearly difficult to formulate. The Greeks sensitively realized that aesthetic proportion is embedded in the figure-ground gestalt of the image and adjusted their golden-section proportions accordingly.

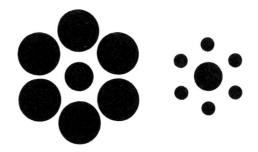

Figure 12-5.

Functional and aesthetic proportions, at their best, tend to be identical. In science and mathematics, for example, an "elegant" experiment or theorem is one that provides the most effect for the least effort; function and beauty are completely integral. Perhaps we develop an appetite for proportional rightness by experiencing it in nature and in the pleasures that beautiful and fit relationships bring to every facet of our life. In any case, visual thinkers who learn to perceive proportional relationships accurately, sensitively, and flexibly will find this ability evident in every aspect of their thinking.

References

1. Smith, I. *Spatial Ability*. Knapp.
2. Doblin, J. *Perspective: A New System for Designers*. Whitney.
3. Le Corbusier. *New World of Space*. Reynal & Hitchcock.

CUES TO FORM AND SPACE

13

Lost in Space

Consider the forms in Figure 13-1. Can you imagine yourself walking this strange staircase that leads endlessly downward?

Or assembling three perfectly square bars into this weirdly warped triangle? Now enter into the spatial worlds depicted in Figure 13-2. In Hogarth's rustic English countryside, relationships between near and far objects are in spatial disarray; in Escher's surrealistic room, even gravity is gone.

False Perspective by Hogarth.

Relativity by M. C. Escher, Collection Escher Foundation, Haags Gemeentemuseum, The Hague.

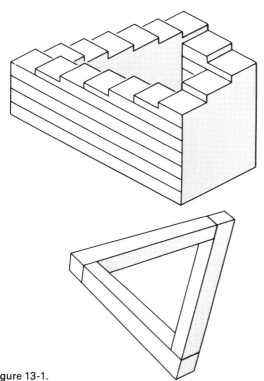

Figure 13-1.

Figure 13-2.

How have the designers of the endless staircase and the warped triangle managed to present you with such bewildering illusions? How have Hogarth and Escher arranged to capture your eye and then to lose it in surrealistic space? They have tricked you by carefully reformulating "cues" by which you normally perceive spatial relationships. This last chapter in the section on seeing will explore these cues.

Spatial Cues Are Learned

Psychologists tell us that spatial perception is not the perception of space as such but of the relationship between objects in space. Further, most psychologists hold that spatial perception is largely learned. Babies, for example, learn that one object is nearer to them than another by gradually discerning the spatial meaning of overlap, illustrated on the left in Figure 13-3. To demonstrate that overlap is a learned spatial cue, readjust your vision to the possibility that the black and white discs in Figure 13-3 are not overlapped but adjoined in the same plane. (The black disc could be incomplete, as shown on the right, and nestled against the white one.) Which is easier, to see these shapes as adjoined or overlapped? Most people find it easier to see them as overlapped, even after another possibility is suggested. Still another possibility exists: did you see the black line as representing a thin loop attached to the black half-moon? Probably not; you have learned to interpret this pattern as representing overlapping shapes.

A more subtle cue to spatial relationship, certainly not learned in the nursery, is atmospheric perspective. The cue of atmospheric perspective is utilized correctly in the Hogarth engraving in Figure 13-2: the rocks, bushes, and trees in the foreground are distinctly darker than those in the hazy distance. When we look at distant objects, we see also the intervening atmosphere; with experience, we learn that this phenomenon signifies deep space.

Three more cues to spatial relationship are illustrated in Figure 13-4: height in plane, relative size, and focus. The cue of height in plane results from the way the eye normally scans near and far objects. The eye usually scans upward as it proceeds into the distance; the far apple, in Figure 13-4, is therefore higher in the visual field. The far apple is also smaller than the near one; the cue of the relative size of objects of known dimension is very important to spatial perception. Further, the far apple is out of focus. The eye cannot focus on near and far objects simultaneously; the mind capitalizes on this physiological limitation to distinguish the relation between objects in space.

Figure 13-4.

Perceptual and Optical Reality

Before exploring additional cues to spatial relationships, consider more carefully how you perceive the form of a single object. Look around you and select a rectan-

Figure 13-3.

gular object (a book or table will do). Are you viewing that object head-on? That is, is your line of vision at right angles to one of the object's rectangular surfaces? Probably not; you are likely viewing most of the objects in your immediate environment obliquely; as a consequence, most of the shapes that you see are distorted. Circles you see as ellipses. Rectangles appear to be trapezoids. Moreover, move your viewpoint and every shape changes. Walk a few paces forward, and some shapes appear to become larger, others appear smaller, and most change proportion. This unreliable visual situation in which solid objects appear to change size and shape I will call "optical reality."

Fortunately for sanity, experience is not primarily optical. In everyday experience, which I will call "perceptual reality," we involuntarily adjust the ever-changing images of optical reality. Psychologists call this perceptual ability "object constancy." By object constancy, you perceive the apples in Figure 13-4 as approximately equal in size, although optically the far apple is only one-third the size of the near one.

Perceptual reality, governed by object constancy, combines what you know with what you see, and that knowing is polysensory. You perceive a chair: polysensory memories merge; you perceive a chair that is solid, pleasant to touch, and soft to sit in. By contrast, optical reality is only visual. The optical image of the same chair is the pattern of color seen by the retina of the eye. *Optical reality is ruled only by geometry; it is the unknowing, mechanical reality of a camera.*

The men photographed in Figure 13-5 are not freaks; they occupy a freak room devised by psychologist Adalbert Ames. As did the visual conjurers whose work was shown earlier, Ames has used optical cues to spatial relationships (especially the perspective cue of convergence) to trick the learned eye into perceiving incorrectly.

13•1 PERCEPTUAL DEMONSTRATION KIT

Psychologists Weintraub and Walker, in collaboration with Brooks/Cole Publishing Company, have devised a simple kit which puts many perceptual phenomena (including Ames' Distorted Room) into experiential form. This kit, which consists partly of cardboard fold-ups, invigorates seeing and also points entertainingly to the limitations of seeing. It is a recommended supplement to the exercises in this chapter. For information on how to obtain a *Perceptual Demonstration Kit,* write to Brooks/Cole Publishing Company, Monterey, California 93940.

The dissonant interplay between the changing images of optical reality and the constant images of perceptual reality provides excellent exercise in seeing. First, experience perceptual reality more consciously, using drawing as a catalyst.

Perceptual Reality and Orthographic Projection

Although we rarely view objects head-on, we always perceive them this way. In everyday perceptual reality, circles are always circles and never ellipses; rectangles are invariably perceived as rectangles, even when viewed obliquely. *The graphic equivalent of seeing things head-on is "orthographic projection,"* a method of drawing commonly used by designers.

The basis of orthographic projection is illustrated in Figure 13-6. An imaginary transparent box has been placed over an object so that the walls of the box are paral-

Figure 13-5.

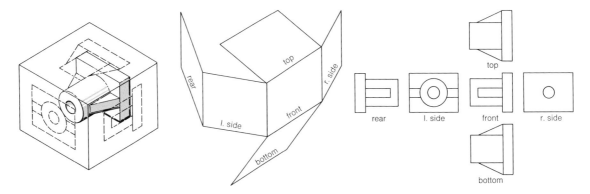

Figure 13-6.

lel to those of the object. Looking perpendicularly through one side of the transparent box, the observer sees one "true shape" of the object. Projecting this true shape of the object onto the corresponding plane of the transparent box is orthographic projection. Other true shapes of the object can be projected onto other planes of the transparent box. Then the planes of the imaginary box are flattened out onto the plane of the page, as also shown, to present multiple views of the object.

In the next exercises (13-2, 13-3), explore perceptual reality of three-dimensional form by means of orthographic projection.

13•2 FRONT, TOP, AND SIDE VIEWS

Select several simple objects. Draw freehand three orthographic views of each object. Refer if necessary to Figure 13-6.

Although we usually see only external surfaces, much reality lies concealed within. Paul Klee noted that "the object grows beyond its appearance through our knowledge of its inner being, through the knowledge that the thing is more than its outward aspects suggest. Man dissects the thing and visualizes its inside with the help of plane sections. . . . This is simple

penetration, to some extent that of a simple knife."[1] In orthographic cross-section, an imaginary plane is sliced through the object to permit visualization of its inner structure, as shown in Figure 13-7.

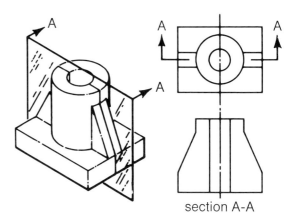

section A-A

Figure 13-7.

13•3 CROSS-SECTION

1. Obtain several varieties of fruits or vegetables that are interesting inside, as are green peppers.

2. Draw orthographic views of each, as in the previous exercise.

3. Then dissect each with a sharp knife. Record what you find inside in two or more orthographic cross-sections.

Optical Reality and Perspective

Just as orthographic projection is the graphic equivalent of perceptual reality, *perspective is the graphic equivalent of optical reality.* Perspective is a comparatively recent perceptual acquisition; there are even today primitive peoples who do not see basic perspective phenomena. R. L. Gregory reports on studies of an isolated tribe that lives in a dense forest: "Such people are interesting, in that they do not experience distant objects, because there are only small clearances in the forest. When they are taken out of their forest and shown distant objects, they see these not as distant but as small."[2] Gregory adds that such people are also initially unable to comprehend perspective drawings and photographs of familiar objects.

The development of perspective in the early Renaissance is a dramatic instance of how human beings learn to see form and space in a new way. As the history of art attests, pre-Renaissance artists were confused by, or ignored entirely, cues to space such as foreshortening and convergence. Today, although the prevalence of perspective and photography diminishes the exhilarating recentering of vision that perspective once delivered, learning to see and draw perspective's spatial cues can still make you aware of space as never before. To experience the optical reality of perspective, try exercise 13-4.

13•4 PERSPECTIVE PICTURE WINDOW

1. Select a window with a view of at least one large rectangular shape, such as a house or street.

2. Look out to the view, with your eyes approximately 18'' from the glass.

3. Fix your head in a stationary position, as you would if you were going to take a long-exposure photograph.

4. Close one eye.

5. With a washable felt-tip marker, trace the outlines of the objects in your view directly on the glass.

The completed drawing on the window pane is, in every respect, a perspective image. The print by Dürer in Figure 13-8 suggests another version of the perspective picture window. By marking a square grid on the glass, and a similar grid on his drawing surface, Dürer's artist was able to draw the perspective image directly on paper.

Although drawing on glass demonstrates the optical reality of perspective, the contribution of graphic perspective is a geometrical procedure for constructing camera-like imagery directly on paper—

Figure 13-8.

without a glass window, and even without an actual model. This procedure is amply treated elsewhere: in exhaustive detail in T. O. McCartney's *Precision Perspective Drawing*,[3] for example, and more simply in Jay Doblin's *Perspective: A New System for Designers*.[4] In the context of this book, we need consider only two basic perspective cues to the perception of deep space: convergence and foreshortening.

Convergence

Seeing and recording the perspective effect of convergence exercises your ability to distinguish optical reality from perceptual reality. The man in the photograph in Figure 13-9, for example, perceives the nearby building as basically rectangular.

However, should he choose to recenter his vision to optical reality, he will see that the parallel lines in the rectangles are not parallel at all. They are converging. The photograph in Figure 13-9 is superimposed with lines to illustrate two important characteristics of convergence. First, receding parallel lines converge to a point, called a vanishing point. Second, when receding parallel lines are horizontal (as they frequently are in our man-made environment), they appear to converge to a point on a horizontal line that coincides with the eye level of the observer. Later, you may want to refer to a perspective textbook to understand the theory that underlies the vanishing point and eye-level locus of vanishing points. More important for now, experience these phenomena for yourself (13-5).

eye level

vanishing point

Figure 13-9.

13•5 CONVERGENCE

1. Convergence is most easily observed in large objects (no smaller than a table). Select several large, horizontal, rectangular objects to draw in perspective.

2. To the best of your ability, make a small, freehand perspective sketch of one of the objects in the center of a sheet of newsprint.

3. Using a different-colored marker, extend all converging horizontal lines in your drawing to a vanishing point (as in Figure 13-9). Then construct a horizontal eye-level line and check to see whether it coincides with your eye level. If your drawing is accurate, it will.

4. Repeat with several other objects.

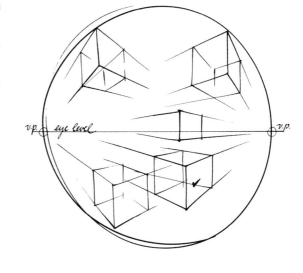

Figure 13-10.

Visual thinkers use perspective primarily to record forms that exist only in their imaginations. Consequently, in the next two exercises (13-6, 13-7) concentrate on seeing convergence not in actual objects but in rectangular solids conceived in your imagination and captured graphically.

Experienced perspective visualizers rarely locate vanishing points on their drawings. Sketching freehand, they record convergence "by eye." In the next exercise (13-7), check your drawings for errors of convergence by using vanishing points afterward.

13•6 TWO-POINT PERSPECTIVE

1. On newsprint, draw the largest possible freehand circle. As shown in Figure 13-10, draw a diameter across the circle and label it "eye level." At each end of this line, draw a vanishing point; these are the two vanishing points of classic two-point perspective.

2. Within the circle, draw a number of horizontal, transparent rectangular solids in two-point perspective. All horizontal edges should be drawn to converge to one of the two vanishing points. All vertical edges should be drawn parallel to each other and perpendicular to the eye-level line.

3. As you draw, imagine that your eye is located on the plane of the line marked "eye level," and that you are looking up or down at each rectangular solid that you draw. Also imagine that each solid is three-dimensional; don't just draw lines; construct a form in space.

13•7 CONVERGENCE ERRORS

1. From your imagination, draw several rectangular solids, freehand, in two-point perspective. View the solids from below and above, and from various angles, but don't use vanishing points.

2. Use a straightedge and different-colored marker to extend *all* of the lines in each completed sketch.

3. Check each drawing for the common convergence errors described on pp. 86–87. Label each error with the abbreviations suggested in the accompanying sketches.

A common beginner's error in depicting convergence is reverse convergence, shown in Figure 13-11, in which receding parallel lines are drawn converging in reverse direction—toward instead of away from the viewer.

Two additional convergence errors are shown in Figure 13-12. Receding parallel lines that fail to converge to a single vanishing point are, of course, "out of convergence." Further, vertical edges should be drawn parallel in two-point perspective; convergence is represented only for edges that are receding from the observer.

In Figure 13-12, one of the verticals is misdrawn; it is askew.

You may be wondering how the distance between vanishing points is determined. In so-called "mechanical perspective," vanishing points are located by geometric procedure; in freehand perspective, vanishing points (if drawn at all) are located by eye. A rule of thumb for the freehand visualizer is that *vanishing points for small objects should be located far apart rela-*

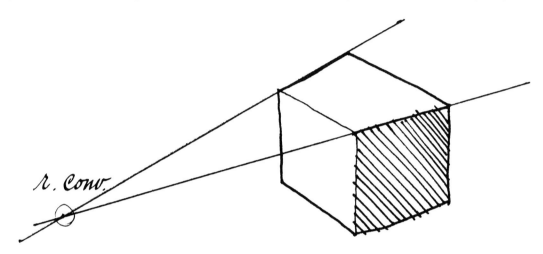

Figure 13-11.

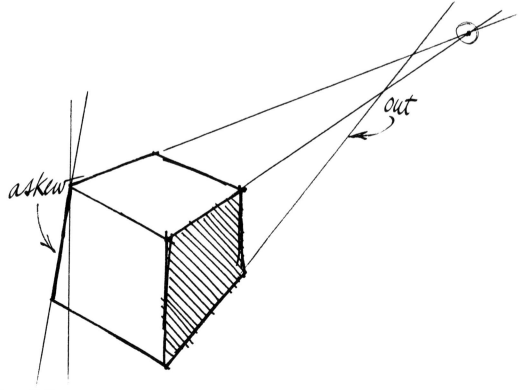

Figure 13-12.

tive to the size of the image, and vanishing points for large objects (such as buildings) should be located relatively close together. In Figure 13-13, the vanishing points are too close, however, causing a misshapen image that resembles a wedge more than a rectangle. Correctly drawn, the corner marked "wedge" should measure not less than 90° on the page. Can you describe why this is so, using an actual rectangle?*

---

*Looking straight down at such a corner, you will see it as a 90° angle. No viewpoint will allow you to see it as less than 90°.

Finally, Figure 13-14 shows a convergence error in which one vanishing point is incorrectly positioned, causing a tilted eye-level line. In two-point perspective, all vertical lines in your drawing should cross the eye-level line at 90°. Notice that the perspective image that results from this error has a peculiar half-misshapen, half-tipped look.

Breaking the hold that perceptual reality has on your vision takes time and effort. The ability to see the optical effect of convergence accurately comes only with repeated practice.

Foreshortening

Close one eye and hold your own hand in front of you, perpendicular to your line of vision. Now bend your fingers toward you until they are almost parallel to your line of vision; your optical image of your fingers is now foreshortened. The cue of foreshortening is important to the perception of the three-dimensionality of form. The next exercise (13-8) will intensify your awareness of this spatial cue.

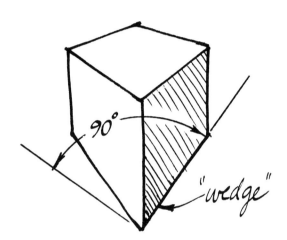

Figure 13-13.

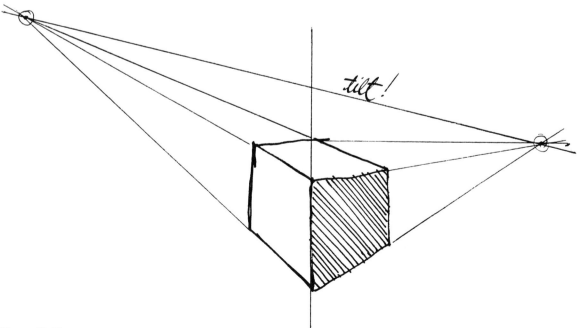

Figure 13-14.

13•8 FORESHORTENING

1. Select a pair of simple identical objects (such as two drinking glasses or two books).

2. Arrange the objects side by side, one almost perpendicular to your line of sight and the other at an angle, as suggested in Figure 13-15.

3. Close one eye. (Binocular vision interferes with seeing perspective cues; perspective, like a camera, is essentially one-eyed.)

4. Draw the two objects in freehand perspective. When you have captured the cue of foreshortening accurately, you will perceive the unlike perspective images as representing identical objects.

5. Repeat, using other objects. See the proportional changes in shape that foreshortening creates. Carefully adjust your drawing until both objects are perceptually identical although optically different.

6. Now draw an object of your own invention in perspective. In adjacent sketches, rotate the object in space (or move your viewpoint) while maintaining the perception of proportion captured in the initial image.

One of the few tricks in perspective drawing involves the drawing of foreshortened circles. In the photograph in Figure 13-15, notice that the foreshortened circles in the upright glass are elliptical in shape. To draw a foreshortened circle in perspective, first draw a perspective square that is the size and is in the plane of the desired circle. Find the center of this perspective square and its midpoints, as shown on the left in Figure 13-16. Now draw a line perpendicular in perspective to the plane of the square and through its center (line A–A in Figure 13-16). Now draw the ellipse: align the minor axis, or shortest dimension, of the ellipse with line A–A; size the ellipse so that it is tangent at the perspective midpoints of the square. With practice, this construction is unnecessary.

Translation

The dissonance between perceptual and optical reality, unconsciously resolved in the brain, causes us to see a stable spatial world of recognizable three-dimensional objects. Again using drawing as a catalyst, make this essential resolution conscious by translating orthographic views into perspective (13-9). This translation will heighten your awareness of the structure of spatial configurations.

13•9 TRANSLATION

1. Sketch several orthographic views of an actual object or an object in your imagination. Be sure to record proportional relationships accurately.

2. Choose a viewpoint from which to draw the object in perspective. A few thumbnail sketches will help you to choose an orientation that reveals maximum information about the form with minimum ambiguity. An unfortunate alignment of edges can, for example, muddle an otherwise well-drawn perspective image.

Figure 13-15.

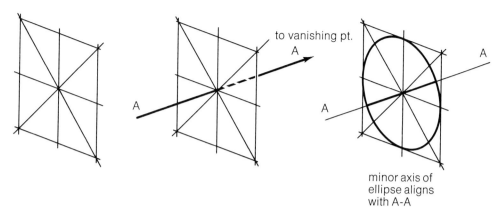

to vanishing pt.

A

A

A

A

minor axis of
ellipse aligns
with A-A

Figure 13-16.

3. With light pencil, draw a transparent perspective skeleton proportioned correctly to contain the object. Within that skeleton, sketch in the structure of the object, including all hidden lines. When sketching in a cylinder, for example, sketch in the hidden ellipse, not just the ellipse that shows. The resulting perspective image is transparent, as illustrated in Figure 13-17.

4. Reinforce the visible lines of the object with a dark marker. Alternatively, overlay tracing paper and trace the visible lines into a new drawing, making corrections where necessary.

Ways to Detect Errors

How can you know when you are seeing and drawing inaccurately? A customary source of feedback is a teacher with a trained eye. But lacking a teacher, you will learn more quickly by frequently using the following recentering devices to detect your own errors:

1. Tack up your drawing and view it from a distance (or through a diminishing lens, available in art stores).

2. Look at your drawing upside down.

3. Look at your drawing in a mirror.

4. If your drawing is on translucent paper, turn it over and look at it through the paper.

Another way to obtain the constant feedback and reinforcement that you need in order to learn quickly is to enlist the eyes of a friend or colleague. It is much easier to see the errors of others than to see one's own. If you can assemble a group, try a group exercise (13-10). It will help you develop your own critical faculties while also helping to develop the perceptual skills of the other members.

13•10 VISUAL RUMOR

Rumors are frequently distorted as they pass from person to person in a group. Make the rumors visual, discuss the distortions that occur in the "visual rumor mill" in a constructive way, and you have a self-teaching group. Optimal group size is four. Time each drawing phase so that the group works at a mutually comfortable pace. Each person begins as follows:

1. *Initial rumor:* from imagination, draw three orthographic views of a simple object on a quarter sheet of newsprint. Label it "A." Pass the drawing on to the next person.

2. *Second rumor:* on a second quarter sheet of newsprint, translate your neighbor's orthographic sketch into a perspective drawing. Label it "B" and pass it on.

3. *Third rumor:* translate "B" into an orthographic sketch, label it "C," and pass it on.

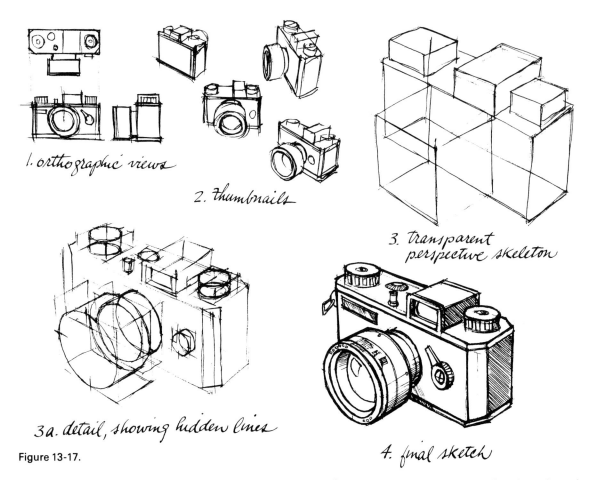

1. *orthographic views*

2. *thumbnails*

3. *transparent perspective skeleton*

3a. *detail, showing hidden lines*

4. *final sketch*

Figure 13-17.

4. *Fourth rumor:* translate "C" into a perspective sketch and label it "D."

5. Place related drawings side by side, in A-B-C-D order. Constructively discuss the errors in the drawings. An alternative game: instead of making orthographic sketches, verbally describe the object for the initial and third rumor.

Color and Shading

The lines that you have been drawing in the preceding exercises do not, of course, exist in nature. Lines in a drawing simply *represent* what you see: they *signify* perceived contrasts in light reflected from objects and their background. While line drawing is appropriate for the visual thinker most of the time, frequent instances occur where line alone is inadequate. Try, for example, to distinguish a sphere from a disc with line alone. Without shading, you cannot depict compound surfaces; nor can you record color, the effect of light and shadow, the spatial drama of figure-ground relationships, and delicate nuances of mood and feeling. To obtain these qualities, you must modify the reflectivity of the paper with color and shading to signify the contrasts in reflected light that you actually see.

Study the idea sketches illustrated in Section V to get a sense of how shading can be added quickly, and in bold patterns, to clarify and unify three-dimensional renderings of form. A word of warning, however: *shading can all too easily become a time-consuming technique well suited to the purposes of communication and poorly suited to the purposes of thinking.* The visual thinker should seek shading that is quick and impressionistic, not photographically realistic.

Return to Pattern-Seeking

The exercises in this and the previous two chapters have tended to focus your eyes on details. After an excursion into details, a return to the whole consolidates and unifies what you have learned. So now return to pattern-seeking, described in Chapter 10; do not allow your attention to be riveted too long by details without returning to the gestalt of the image. One way to return your vision to the patterning mode is suggested by the next exercise (13-11). Notice how *analytically acquired perceptions of perspective become imaginative material for the pattern-seeking phenomenon of projection.*

13•11 PERSPECTIVE IN PATTERNS

1. With a light gray felt-tip marker, *scribble* a perspective image of an object. Build up the image gradually, as you would a lump of clay, with loose scribble lines.

2. With a darker marker, such as a black ballpoint pen, seek a more accurate perspective image in the scribbles (see exercise 10-5/Da Vinci's Device).

Seeing Is Limited

Although seeing allows endless innovation, it is also limited. Indeed, artists have invented new ways to see in order to expand vision's boundaries. As we have discussed, Renaissance artists revolutionized the way we see deep space. Subsequent artists, rebelling against the "materialistic objectivity," fixed viewpoint, and static nature of the perspective image, reformed seeing into new modes. Cézanne, for example, invented a multi-viewpoint perspective. Other artists invented "abstract" ways of seeing that divest imagery of its representational content and center visual attention on aesthetic-sensory qualities invested in color, texture, line, shape, and space. Innovators in science and technology have also extended the visual sense. Through powerful microscopes and telescopes, we now are able to see new landscapes in microstructures and in the cosmos. By photography and television, we now visually experience world events as they happen; and by a broadcast from outer space, we visit the moon and even look back to the tiny ball that is our planet Earth.

However, just as this section on experiences in seeing is limited and cannot encompass all varieties of seeing, so seeing is also limited. The visual sense operates within a finite range. Many important phenomena elude our visual sense. Even with our most advanced technological extensions of the eye, we cannot see objects that are extremely small, such as atomic particles. Nor can we see events that are extremely large and complex, such as urban patterns of social interaction. We cannot closely observe happenings that are very far away from our organs of vision, such as distant quasars. Nor can we glimpse the inner structure and workings of many things, including the inner core of our planet and the functioning of our brain. Our visual sense is also insensitive to most energy phenomena: we see a limited range of light frequencies and cannot see occurrences such as forces emanating from the poles of a magnet. Further, the eye is bound in time and space: we cannot see the past or the future; eyes that are here-and-now cannot also be there-and-then. By definition, ESP phenomena occur outside the reach of the visual sense.

More important, *what we see is limited by what we know. And when our knowledge is incorrect, we see illusions.* The poet William Blake advised, "Consume the sensory perceptual illusions in the fires of imagination." Einstein exemplified this advice: by his imagination, he exposed fallacies in the Newtonian model of the universe that had distorted the perceptions of generations of physicists. Similarly, Jesus, Buddha, Darwin, and Freud, by their imaginative insights, drastically changed the way we see ourselves.

Seeing is a precious gift, but a limited one that cannot be divorced from imagination. Visual discovery, in the scientific laboratory, on the artist's canvas, and in human relationships, occurs in relation to a form of inner vision that is far freer of mortal limitation than is sensory perception. The task of the next section, "Imagining," is to lead you to experience more fully the contribution of imagination to the interactive process (seeing/imagining/drawing) of visual thinking.

References

1. Klee, P. *The Thinking Eye.* Wittenborn.
2. Gregory, R. *Eye and Brain.* McGraw-Hill.
3. McCartney, T. *Precision Perspective Drawing.* McGraw-Hill.
4. Doblin, J. *Perspective: A New System for Designers.* Whitney.

Recommended Reading

Obtaining skill in orthographic and perspective projection is not only a vigorous discipline in seeing but also an invaluable aid to imagining and idea-sketching the spatial structure of visual ideas. If you want to learn more about orthographic projection, you will find that any of the standard architectual or engineering graphics texts cover this subject thoroughly. If you want more instruction in *freehand* perspective, refer to E. W. Watson's *Perspective for Sketchers* (Reinhold).

The approach to the education of vision taken in this section can be complemented by many other approaches. The *Vision and Value Series,* edited by Gyorgy Kepes and published by Braziller, is an especially rich source of alternatives.

IMAGINING

IV

The following six chapters are devoted to strengthening the organ of inner vision, the mind's eye. The mind's eye, spontaneously active in dreaming, can also be consciously directed. Unlike the sensory eye, which is bound to the here-and-now, the mind's eye can travel in space and time to the there-and-then, can entertain fantasy, can form, probe, and manipulate structures and abstract ideas, can obtain insight into realities that have not yet been seen; and can foresee future consequences of present plans. "The faculty of imagination," wrote Dugald Stewart, "is the principal source of human improvement."

THE MIND'S EYE

14

What Is Imagination?

Human imagination enables us to transcend mortal limitations of space and time, to experience what was, what can be, what can never be; it opens vistas that are not available to the senses. "By his imagination," writes Frank Barron, "man makes new universes which are 'nearer to the heart's desire.' The sorcery and charm of imagination, and the power it gives to the individual to transform his world into a new world of order and delight, makes it one of the most treasured of all human capacities."[1]

Before exploring the inner imagery of the mind's eye and the workings of visual imagination, however, consider imagination first in a larger context. The fish does not know he is wet. Similarly, imagination so permeates human experience that many people frequently claim they have no imagination. Imagination is more than the power to be creative. Imagination is all that you have ever learned or experienced; it is central to your every perception and act. Experience this in exercise 14-1.

14•1 WHERE ARE YOU?

Starting with present location and working outward, locate yourself spatially (noting how your imagination enables you to do so). First, what is behind you? You need not look around: in your imagination, experience the furniture that you know is there. Now go through the wall in front of you, into the next building, as far as you can in that one direction. Next, without looking down, experience what is underneath you: travel in your imagination down through the floor to the bowels of the earth. Finally, locate yourself in relation to oceans and land masses and picture your earthly locus in relation to the solar system.

Did you say that you don't have imagination? You'd be literally lost without it. Furthermore, your sweeping awareness of space is matched by your capacity to relate yourself to large spans of time. You know not only *where* you are, but *when* you are. Your sense of history as well as your ability to plan for tomorrow are the work of your imagination.

Whenever you answer the question "How?" or "Why?" you are also exercising your imagination. How do flowers reproduce? Why are you reading this book? The ability to create imaginative bridges between cause and effect, almost entirely undeveloped in animals, is a marked faculty of human imagination.

Another ability of imagination that you possess, and animals do not, is self-awareness. You have an image of yourself, or more correctly, a number of self-images. For instance, in a classroom you may see yourself as a diligent student; at a party you will likely abandon this self-image for another. You also project images on other people: you may see an individual as a policeman, for example, but not as a husband and father. To become aware of the pervasive role of imagination in personal and interpersonal behavior is to begin to understand the games people play.

Imagination also rules what you choose to see or ignore, to like or dislike. *Every moment you match your immediate experience with previous experiences stored in image form.* The pleasant and unpleasant feelings that accompany this ongoing subconscious matching process then largely dictate what you choose to attend, and how. This pervasive effect of imagination accounts, for example, for differences in the way two people see the same event (a dentist and a psychologist see the same smile differently) as well as for differences in the way two people choose from the same alternatives (make mine vanilla). Return to Chapter 8, especially to the discussion of Figure 8-1, for a more extensive treatment of imagination's role in the act of perception.

Finally, no one who dreams can correctly claim to be unimaginative. Whether you remember your dreams or not, you are imaginative, often profoundly so, for about an hour and a half every night.

Imagination, then, is not limited to occasional leaps of fancy or rare moments of creative insight. Imagination, the total of all you know and have experienced, is at the center of your being. Everyone possesses imagination. People who claim they have no imagination are simply confessing their lack of awareness.

Real or Illusory?

Imagination can be reality-oriented: by imagination, we correctly pictured the earth's roundness long before astronauts showed us the confirming photograph. But imagination can also be illusory: the child who plays with an imaginary friend may need an actual friend, but has an illusion. *We correctly distrust the illusory capacity of imagination, but we need not allow this distrust to become prejudice against imagination generally.* Undiscriminating prejudice against imaginative expression is exemplified by parents who teach their children to distinguish the illusory from the real by demanding that they "stop imagining things!"

Why is imagination treated by many as if it were forbidden fruit? Flights of imagination are popularly associated with insanity; the physical inactivity that normally accompanies imagining is easily mistaken for that bane of work-oriented cultures, laziness; "idle daydreaming" is considered to be escapist; imagination often leads to maladjusted behavior (in other words, behavior that does not conform to "social reality").

There is truth to each of these claims against imagination. Entire cultures, not just individuals, *are* sometimes dominated by insane imaginings. Imagination *is* nurtured by cessation of outward activity and stimulation. And creative imagination, especially of the most profound sort, *is* socially disruptive. Indeed, no force yet discovered is more powerful than imagination. Like nuclear energy, the power of imagination can be used creatively or destructively, as we choose to direct it—*if* we are able to choose.

Unfortunately, our ability to choose is severely diminished by those who educate us to "stop imagining things" and to "face up to reality." What reality? Frequently we are taught to look to authority for truth: the teacher, the priest, the political leader, the scientist. History amply proves that the reality held by authority figures is by no means infallible. Then trust your senses: "seeing is believing." Every course in introductory psychology demonstrates how the senses can be tricked. Then look to consensus: "Ten million Frenchmen can't be wrong." Or look to your feelings: does it feel right? In the final analysis, all of these and all other roads to reality come upon a decisive detour. Imagination lies between the stimulus and the perception.

The power to "capture the imagination" often describes the ability of leaders to seduce and victimize a group of followers. Consider the millions of people whose imaginations were captured by Hitler's version of reality. Consider the millions whose lives are today molded by myths promulgated by Madison Avenue and the mass media. How can the proverbial ten million Frenchmen be right when their imaginations have been programmed with distorted, biased, and even blatantly untruthful images? Only by being aware of imagination can we know its illusory powers; the person whose imagination is active and trained can chart a course through illusion toward reality. By contrast, people who are passively unaware of their imaginations are illusion's easy victim.

Start Imagining Things

Imagination denied is not reality enthroned. Rather than "stop imagining things," *look inward, become aware of your imagination, and learn to control it productively.* If you are already aware of inner imagery but are unable to control it (if you are prone to drift in cyclic and unproductive daydreams, for example), then learn to extend your awareness into other modes of imagery, and again, learn to direct your imagination toward productive ends.

In the previous section, on seeing, you focused your attention outward; in this section, focus it inward. Consciously alternating between outer and inner awareness brings the differences and relative values of seeing and imagining into sharper definition.

What Is a Mental Image?

A mental image, unlike a perceptual image, can occur without sensory stimulation, with the eyes closed. Mental images can be experienced in every sensory mode: images may be "seen" in the mind's eye, for example, or "heard" in the mind's ear.

Although virtually everyone experiences mental imagery nightly, in their dreams, the ability to experience clear mental images while awake varies from one person to another. Approximately one-sixth of a normal group of people "do not use visual images in their thinking unless obliged to do so. Even then, their mind's eye is almost blind." Another sixth see vivid inner imagery, and the remaining two-thirds "can evoke satisfactory visual patterns when necessary."[2] A similar in-

equality is evident in the other modes of sensory imagination.

Although inherited ability is one factor at work here, those people who see little or no inner imagery should not be too quick to consider themselves genetically shortchanged. It is my contention that *almost everyone can learn to experience and use some form of mind's-eye imagery.* What needs to be compensated for in a majority of educated people is a default in their education. Failure to pass a physical fitness test is usually evidence of lack of exercise, not lack of muscular inheritance. The same goes for ability to evoke mental imagery.

Contemporary education seriously neglects inner imagery in two ways. First, it fails to make people aware of their inner imagery. Second, it affords little opportunity for them to develop this inner resource. Bats and fish, forever confined to dark places, evolve blind eyes. Likewise, the human mind's eye must be used, or it too goes blind.

Fortunately, the mind's eye can be reopened and revived. One simple method, seeing after-images, is used by psychologists whose subjects are required to use mind's-eye imagery in experiments. Although an after-image is a retinal effect and not a mental image of the sort seen in dreams, it has the virtue of being available to everyone with eyesight. People who claim to see no inner imagery can usually see an after-image, and from this experience of what a mental picture is like they can begin to understand the condition of inner attention that enables the mind's eye to see true mental images. Experience a vivid after-image in experiment 14-2.

14.2 AFTER-IMAGE

1. From bright green paper, cut out the silhouette of an elephant. In the center of it, place a black dot. Put the silhouette on a dark background, in a well-illuminated place.

2. Without wavering, stare at the black dot while slowly counting to ten.

3. Look up to a plain wall and see a large pink elephant. Or close your eyes and see a small one.*

Alternatively:
1. Look at the center of the Shri Yantra Mandala in Figure 14-1.

Figure 14-1.

2. "Breathe easily, *in* through the nose—flaring the nostrils. *Out* through the mouth—6 times—slowly.

3. "Close your eyes and focus them up into the middle of your forehead. The emerging after-image resembles the Central Sun. Use your will to brighten the colors."[3]

Be sure to repeat the after-image experiment until you are able to see one. Alan Richardson points out that some people need practice before they can see after-images, especially if they have never before paid attention to internal channels of stimulation.[4]

After being sure that you have seen an after-image, you can take a next step toward increased awareness of your inner imagery by attending whatever patterns of color appear to your mind's eye when

*Try flashing your green elephant or a similar image with a T-scope on a darkened screen (see Chapter 10). In this method, the after-image will often be the same color as the projected stimulus.

your eyes are closed in a darkened room (14-3). These patterns are frequently so muted that they are easily overlooked. Psychologist Peter McKellar notes that even Galton, a practiced introspectionist, was surprised to observe that his eyes-closed vision was not uniformly black, as he had previously believed, but a kaleidoscopic change of patterns and forms. McKellar elaborates further that "these luminous visual occurrences have been given various names including 'luminous dust,' 'idioretinal light,' 'phosphene,' and 'Eigenlicht.'"[5]

14•3 LUMINOUS DUST

Close your eyes in a pitch-dark room. Tightly "squinch" your eyelids, but *don't* rub them. Become aware of the luminous patterns, however vague, that appear before your mind's eye.

Although science has not yet discovered how and where mental images are formed, evidence indicates that they depend on processes both within the brain and within the retina of the eye. One hypothesis is that the eye supplies amorphous phosphene patterns that the brain organizes into meaningful mental pictures. In Chapter 10 you experienced the phenomenon of projection, in which meaningful images are seen in amorphous cloud forms or swirls of fingerpaint. Similarly, meaningful mental pictures can be projected on the phosphene patterns that occur before your mind's eye (14-4).

14•4 PHOSPHENE PROJECTIONS

Repeat the previous internal viewing of luminous dust, this time finding pictures suggested in the phosphene patterns. (See also exercise 10-5, Da Vinci's Device.)

Conditions That Foster Inner Imagery

If your mind's eye has not yet been aroused, do not be impatient or disappointed. The chapters that follow present a variety of exercises; one of these may prove to be your mind's-eye opener. Practice is also essential: if your organ of inner vision has been long dormant, repeated effort may be necessary to reawaken it.

Further, inner imagery is fostered by the following conditions:

1. A quiet environment. Inner stimuli, often fragile, are easier to attend in an environment in which external stimuli (such as distracting noises or interruptions) are absent.

2. Motivation. Attention follows interest; you will more likely see mental pictures when you are motivated to do so (but not overmotivated).

3. Relaxed attention. Review Chapter 6; a state of relaxed attention is particularly important to inner imagery. Without a certain degree of relaxation, imagery is suppressed. Without a modicum of attention, imagery is random. What is needed, particularly in the eye muscles, is "optimal tonus."

4. Find the locus of your imagery. For some people, inner imagery is "out there," in front of the eyes. For others, mental pictures are experienced in the middle of the forehead or in the back of the head. Some people see mental pictures more readily with their eyes open; others with eyes closed. Experimentation will help you to find the locus of your imagery.

Other means are also known to elicit visual imagery. Hypnotic and psychedelic experiences, for example, are frequently accompanied by enhanced inner imagery. Photic stimulation—looking into a flashing strobe light synchronized with the alpha rhythms of the brain—commonly elicits mental pictures. Sensory deprivation experiments, in which the subject is deprived of all sensory input for a long period of time, often cause visual hallucinations. In this book, I will present other means to elicit imagery, means that do not require expert supervision and are well within the safe grasp of the reader.

Clarity and Completeness

Many people who claim that they do not see mind's-eye imagery may actually be expecting too much. Writes Charles Osgood: "Images are generally less clear, stable, and saturated than perceptions."[6] And Sir Frederick Bartlett: "A visual image may be of every stage of completeness, from the most fleeting fragment to the most literal reinstatement."[7]

Indeed, *vivid and clear imagery is frequently not desired in visual thinking*. Mental operations that involve abstraction, flexible manipulation, and creative synthesis can actually be obstructed by detailed mental pictures. Visual memory, on the other hand, is generally facilitated by clear imagery, as is the visualization of concrete ideas.

Experience 14-5—an adaptation from a psychological test—will permit you to assess the vividness of your visual imagery at the present time. Subsequent experiences are intended to help you intensify and clarify your imagery. In the next chapter, for example, you will find that imagery is intensified when all sensory modes are brought to play. Test yourself now. After you have exercised your mind's eye-ear-nose and so on for several months, test yourself again.

14·5 CLARITY OF MENTAL IMAGERY

Translate each of the following descriptions into a mental image. As you do, rate its clarity according to the following scale:

C = Clear
V = Vague, but recognizable
N = No image at all

Can you visually imagine:

1. A familiar face.
2. A galloping horse.
3. A rosebud.
4. A body of water at sunset.
5. Your bedroom.
6. The characteristic walk of a friend.
7. A table laden with food.
8. A changing stoplight.
9. The moon through clouds.
10. A newspaper headline.

The following descriptions are intended to evoke other modes of sensory imagery:

1. The sound of rain on the roof.
2. A bird twittering.
3. The voice of a friend.
4. Children laughing at play.
5. Thunder.
6. The feel of soft fur.
7. The prick of a pin.
8. A cold shower.
9. An itch.
10. A soft breeze on your face.
11. The muscular sensation of running.
12. Of sitting down in a comfortable chair.
13. Of kicking a can.
14. Of drawing a circle on paper.
15. Of reaching toward a high shelf.
16. The taste of a lemon.
17. The taste of black pepper.
18. The taste of salt.
19. The taste of toothpaste.
20. The taste of a green onion.
21. The smell of bacon frying.
22. The smell of a gardenia.
23. The smell of perspiration.
24. The smell of burning leaves.
25. The smell of gasoline.
26. The sensation of hunger.
27. Of extreme fatigue.
28. Of a cough.
29. Of coming awake.
30. Of radiant well-being.

Controllability

The ability to control imagery is far more important to visual thinking than the ability to conjure clear mental pictures. In the next experience (14-6), you will learn how well you are currently able to control the activity of your mind's eye. As before, test yourself again several months from now. Exercises such as those in "directed fantasy" (Chapter 17) should improve your ability to control your imagination.

14•6 CONTROL OF MENTAL IMAGERY

Taking your time, translate each of the following descriptions into a mental image. As you do, rate your ability to control the image according to the following scale:

C = Controlled the image well
U = Unsure
N = Not able to control the image

1. A rosebud, very slowly blooming.
2. An airplane propeller, rotating clockwise as you face the airplane, then rotating counterclockwise.
3. A stone dropped into a quiet pond; concentric ripples forming and expanding outward.
4. A gray kitten that turns blue, then green, then purple.
5. A red apple hanging on a tree and then re-

gressing in time, becoming greener, smaller, eventually transforming into an apple blossom.

6. This book flying away, high into the blue sky, finally disappearing.

7. A car crashing head-on into a giant feather pillow.

8. The previous image in reverse motion.

9. A table gently floating to the ceiling, unaided, and turning upside down on the way.

10. Your shoe coming apart in slow motion, and each piece drifting away into space.

11. Your chair coming alive and carrying you into the next room.

12. An orange being cut into five equal pieces and the pieces being arranged in three different patterns.

Do not be disappointed if you did poorly on these tests. Just as you would not expect to pass any test that covered material you have not used for years, do not expect sudden mastery in sensory imagination if you have not been using it.

Instead, practice the exercises for the mind's eye in the next few chapters.

References

1. Barron, F. "The Psychology of Imagination," in *A Source Book for Creative Thinking* (edited by S. Parnes and H. Harding). Scribner's.
2. Walter, W. *The Living Brain.* Norton.
3. Strauss, R. *How to Win Games and Influence Destiny: Book 11.* Gryphon House.
4. Richardson, A. *Mental Imagery.* Springer.
5. McKellar, P. *Imagination and Thinking.* Basic Books.
6. Osgood, C. *Method and Theory in Experimental Psychology.* Oxford University Press.
7. Bartlett, F. *Remembering.* Cambridge University Press.

Recommended Reading

Mardi J. Horowitz's *Image Formation and Cognition* (Appleton-Century-Crofts) is a research-oriented contribution to the literature on imagery, with material that ranges from clinical case studies to neurobiological research. Robert Sommers's *The Mind's Eye* (Delta) is a thorough treatment of the subject written in personal style that will engage the lay reader. Mike and Nancy Samuels's *Seeing with the Mind's Eye* (Random House) is a delightfully written and beautifully illustrated book that captures the rich potential of mental imagery, especially in the realm of psychosomatic healing.

VISUAL RECALL

15

new ideas. Your ability to remember past sensory experience was tested in the preceding chapter. If this ability was less than you desired, this chapter may help you to improve it.

Visual recall can be short-term or long-term. Briefly experience each of these forms in the next two exercises (15-1, 15-2).

15•1 MEMORY FOR DOODLES

1. On a sheet of newsprint, scribble a simple, nonrepresentational doodle.

2. Cover the doodle with your hand and draw your short-term memory of it.

Memory and Thinking

Memory is not only useful for finding misplaced objects, recognizing long unseen friends, recollecting facts, and remembering the good old days. Memory is essential to thinking. "Thinking," writes Sir Frederick Bartlett,

> is the utilization of the past in the solution of difficulties set by the present. . . . By the aid of the image, and particularly of the visual image . . . a man can take out of its setting something that happened a year ago, reinstate it with much if not all of its individuality unimpaired, combine it with something that happened yesterday, and use them both to help him to solve a problem with which he is confronted today.[1]

Recall

The mental operation of visual memory, as suggested in Chapter 3, is but one of a number of visual-spatial operations. In this chapter, I will call memory of visual images "visual recall" to distinguish it from thinking operations that combine and transform visual memory images into

---

15•2 CHILDHOOD HOME

From long-term memory, draw a front view and floor plan of a childhood home.

Forgetting

A striking characteristic of long-term visual recall, as you may have just experienced, is how much is apparently forgotten. Schachtel observes that most adult memories of childhood border on total amnesia. Remembrances of the good old days are commonly infused with wishful fantasy. Looking back, we confront gaping holes in our memory that we patch up as best we can.

Forgetting does not mean that past experience is lost, however. Scientific experiments reveal that people possess images in memory that are far more accurate than the ones they are able to recall consciously. "Hypermnesia" is the psychological term for extraordinary acuity of recall. Lawrence Kubie describes an experiment involving hypermnesia under hypnosis in which a subject allowed to view a strange room for a few minutes would subsequently be able to list 20 or 30 of the items he had seen. Under hypnosis, however, he would remember 200![2] Wilder Penfield, in a series of remarkable experiments involving direct electrical stimulation of the brain tissue, elicited detailed and vivid memory playbacks that had the force of actual perceptions.

Hypnosis and electrical stimulation of the brain, admittedly unusual methods of memory retrieval, strongly suggest that *forgetting is not a failure of storage so much as a failure of retrieval.* By your own experience, however, you may be aware of other important factors involved in forgetting. Do you, for example, tend to quickly forget experiences that bore you and to remember more readily experiences that engage your interest?

Eidetic Imagery

Although most adults experience clear visual imagery only in occasional dreams, a rare class of people are able consciously to recall images that are exceptionally clear and accurate. These vivid mental pictures are called "eidetic images."

What is an eidetic image like? Eidetikers report memory images that are very much like perceptual images: they are clear, bright, and detailed, and can be scanned much as one scans an actual scene or a photograph. The eidetic image, often unfaded years after the initial perception, is not truly "photographic," however. Like perception, eidetic recall is subject to inaccuracy; the eidetic image is scanned selectively, according to the viewer's needs and interests.

Few would deny the value of eidetic recall. McKellar reports the introspection of an eidetic student: "I was fifteen. During an exam, I 'saw' my chemistry book. I mentally opened it, turned over the pages and 'copied' the nitric acid diagram into my exam paper."[3] In addition to its obvious utility in memory tasks, eidetic imagery can also play a role in productive

thinking. Tesla, as reported in Chapter 1, thought by eidetic images to invent the A-C generator. Eidetikers are commonly able to manipulate and to fantasize their vivid imagery.

What is known about this phenomenon of acute visual recall? Studies of eidetic imagery are rare and contradictory—understandably enough, since investigators of inner imagery have no way of observing what their subjects are imagining. But though the studies are inconclusive, as Alan Richardson observes, they still tend to support the traditional view that "eidetic imagery is part of a more general mode of concrete functioning which in the normally developing Western child disappears as he enters on his high school education." As schooling advances, "language is used more and more to compress, to represent, and to express experience." Language labeling is presented in the classroom as mentally economical, Richardson continues: "To re-see, re-hear, and re-feel the experience is uneconomical. Under these conditions it is perhaps not so surprising that the ability to use eidetic imagery in those who once possessed it begins to wither away from lack of use."[4]

Gardner Murphy, considering atrophy of mental functioning from a neurological viewpoint, suggests that "cultivation of imagery is in part the physiological cultivation of cortical functions."[5] Uncultivated cortical functions diminish in strength when they are not used, much as muscles weaken when they are not exercised. Conversely, *the imagery center of the brain is responsive to exercise.* Imagery that has been allowed to wither away from lack of use can at least be partially regained by "visual calisthenics."

Memory and Seeing

"People who look at things without seeing them," writes Frederick Perls, "will experience the same deficiency when . . . calling up mental pictures, while those who . . . look at things squarely and with recognition will have an equally alert internal eye."[6] By this rule, a primary by-product of the experiences in the previous section on seeing should be the enhancement of visual imagination. The converse is also important to note: people who have acute visual memories usually see more fully than those who do not. *Memory and seeing are mutually reinforcing.* They enhance each other, as you will see in the next exercise in short-term visual recall (15-3), suggested by Aldous Huxley in *The Art of Seeing*.

15•3 FLASHING

---

1. Look at a scene in your environment—a wall, for example. See it as an overall pattern (Chapter 10) and also see it analytically (Chapter 11). Shift your attention from point to point; follow outlines and count the salient features.

2. "Then close your eyes, 'let go,' and conjure up the clearest possible image of what you have just seen.

3. "Re-open the eyes, compare this image with reality, and repeat the process of analytical looking.

4. "Close the eyes, and once more evoke the memory image of what you have seen. After a few repetitions, there will be an improvement in the clarity and accuracy both of the memory image and the visual image recorded when the eyes are open."[7]

The experience of "flashing" utilizes two important principles. First, it asks you to see attentively and analytically. Ulric Neisser observes: "To construct something attentively is to see it clearly. Such objects can then be remembered; that is, they can be reconstructed as visual images." Second, flashing utilizes a form of short-term memory that Neisser calls "iconic memory." The iconic image is available to virtually all of us if we focus our attention inward directly after a flash of seeing; it is a kind of mental snapshot available "for a second or so after a single brief exposure."[8]

Huxley points out that the flashing experience encourages mind and eyes to function together, instead of functioning apart as they do when the mind is allowed to daydream while the eyes see "just enough to avoid running into trees or under buses." Elaborating on this notion, Huxley advises daydreamers to close their eyes so that sensory and imaginative visual functions cannot operate divisively. And he counsels thinkers who work with their eyes open to use their eyes

> to do something relevant to the intellectual processes going on within the mind. For example, write notes which the eyes can read, or draw diagrams for them to study. Alternatively, if the eyes are kept closed . . . let the inward eye travel over imaginary words, diagrams, or other constructions relevant to the thought process which is taking place.

An experience that further utilizes Huxley's advice is memory drawing (15-4).

15•4 MEMORY DRAWING

---

1. As in the previous exercise, "flash" an object or scene several times.

2. Then look only to your paper and remember, by drawing, the memory image of what you've seen. Don't look back: draw strictly from memory.

3. Now compare the graphic image with the perceptual one.

Cognitive Structures

Now let us consider long-term memory. Are old mental pictures stored in the brain much as photographs are pasted into an album? According to recent psychological theory, mental pictures are not stored at all. What is stored is the *process* by which each image was initially perceived. Psychologists call the receptacle for this process a "cognitive structure." Cognitive structures are not like a cabinet of drawers; they are actively integral with memory. An analogy to videotape may help to clarify.

Videotape is the coated plastic ribbon on which television images are stored. Unlike camera film, videotape does not store actual pictures; instead, it stores an electronic process somewhat similar to visual scanning. Videotape reproduces a television picture only when it is played back on a videotape machine. In the playback, the machine rescans the moving videotape. When this rescanning operation is exactly the same as the original scanning, an accurate remembering occurs. When the rescanning process is not synchronous with the initial scanning, however, the result is an imperfect or blank image.

Cognitive structures are analogous to videotape in that they store a process, not a picture, and are accurately remembered only when the initial storing process is correctly reactivated. I am tempted to carry the analogy somewhat further by imagining a videotape library in which similar tapes are stored together, just as similar memories, or cognitive structures, are stored together in the neural networks of the brain. But at this point the analogy to videotape breaks down. A videotape library stores tapes separately; information on one tape cannot merge with information on another. Not so with cognitive structures; memories easily merge. Further, human playback is not machine-like. Humans do not rescan their cognitive structures with routine precision but with needful selectivity that varies from day to day. This variability makes humans creative in a way that machines are not. It also makes humans less reliable at remembering than machines.

Which is not to say that human memory is not indexed. Cognitive structures are indexed and cross-indexed in the brain in incredibly complex fashion. The most commonly used indexing scheme for cognitive structures is language. As you read, listen, or talk, you use the indexing scheme of your mind with incredible rapidity. Earlier, upon reading the words "childhood home," for example, you immediately had access to at least some portion of the related cognitive structure. Had I been able to give you an evocative and concrete verbal description of your childhood home, you would of course have had access to even more.

Words can influence visual recall both positively and negatively. By the stereotyped use of words, "the memories of the majority of people come to resemble increasingly the stereotyped answers to a questionnaire, in which life consists of time and place of birth, religious denomination, residence, educational degrees, job, marriage, number and birth dates of children, income, sickness, and death."[9] Several nonverbal exercises in Chapter 8 are intended to help you break the negative hold of stereotyped labeling on your "indexing" of experience. Exercise 11-4, Verbal Seeing, suggests how language can be used positively toward the same objective; see also the exercises in "relabeling" in Chapter 8. Exercise 15-5 suggests how you can use reading to exercise and enliven vivid sensory recall—another positive use of language.

15·5 MIND'S-EYE READING

Whenever you read, simultaneously translate the verbal description into full polysensory imagery. For example, when reading a news item, visualize people, locale, and sequence of events; use your other senses as well: hear sounds, smell odors, and so on. A well-known speed-reading course advises students to scan the words of novels while simultaneously seeing the plot unfold in an internal cinema of sensory imagination. Image-laden poetry is an especially rich vehicle for mind's-eye reading.

Relaxed, Multimodal Retrieval

As remarked earlier in this chapter, experiments in hypnosis and electrical brain stimulation suggest by their startling results that accurate recall is strongly related to retrieval methods. The concept of cognitive structures also lends credence to this view, as experiences 15-6, 15-7, and 15-8 demonstrate.

15·6 APPLE I

1. Close your eyes and recall an apple in your mind's eye. (Don't read further until you've tried to visualize the apple.)

2. After a minute or so, open your eyes and ask yourself: "Did I see a colored apple? Was it a specific apple? What was the apple resting on?"

Most people, when attempting to recall an apple, either experience blank or inobedient imagery or a stereotyped red apple that floats in space. Now consider the rationale for a retrieval method that will enable you to recall more vivid and complete memory images. I will call this method "relaxed, multimodal retrieval."

Why relaxed retrieval? When consciousness is relaxed, as it is in hypnosis, for example, long-term memories are more readily recalled. Indeed, a common hypnosis demonstration is "age regression," in which the subject is returned to an experience he or she had as a child or infant. Vividly reliving the memory, the subject frequently performs astonishing tasks of recall. For example, an adult raised in Germany until he was nine and then brought to the United States was regressed to childhood; in the regressed hypnotic state, he remembered his long-forgotten German tongue and, in fact, refused to talk to the hypnotist in English.

The importance of relaxed attention (see Chapter 6) to visual recall can be explained in terms of cognitive structures. Unlike videotape, cognitive structures encode information in every sensory mode and in the mode of feeling. Much of this intersensory and feeling input is assimilated subconsciously. When you last munched an apple, for example, were you fully conscious of its nuances of color,

flavor, scent, coolness, crispness, and texture? Likely not. More probably, you were talking to someone or thinking of something else. If my assumption is correct, your image of an apple in the previous exercise was as lacking in sensory detail as your usual conscious experience of apples. As with videotape, cognitive structures can be replayed accurately only in the same mode that they were recorded. Thus you must relax consciousness to replay memories partially recorded subconsciously.

Memory retrieval is also enhanced when recall is multimodal—that is, when all sensory modes of imagination and the mode of feeling are called into the playback. Thus Perls writes that imagining involves

> more than just visualizing pictures. If you visualize a landscape you can [envision] all the details: the trees, the shadows, the grazing cattle, the fragrant flowers. But you must do more. You must walk in it, climb the trees, dig the rich brown earth, smell the blossoms, sit on the shadowed grass, listen to the birds singing, throw stones in the stream, watch the bees about their busyness! . . . This sensomotoric approach, especially that of touching . . . will develop your sense of actuality and will bring about that eidetic memory (identity of perception and visualization) which in dreams themselves is always present.[6]

To Perls's advice, I would add that the mood or feeling of the imaginative experience, if not evoked naturally by the sensory imagery, should also be elicited.

Before trying the next experiment, find a quiet place, sit down, and relax. Or more correctly, relax attentively. Patient practice will teach you the balance of relaxation and attention that allows you to elicit clear inner imagery. Now try a relaxed, multimodal approach to visual recall (15-7).

15·7 APPLE II

1. Close your eyes and relax; direct your attention inward.

2. Imagine yourself in a familiar setting in which you would enjoy eating an apple. Relaxedly attend the sensory detail and mood of this place.

3. Now imagine that in your hand you have a delicious, crisp apple. Feel the apple's coolness; its weight; its firmness; its round volume; its waxy smoothness. Explore its stem. Visually examine details: see bruises; the way sunlight sparkles on the facets of the apple form; the way the skin reflects a pattern of streaks and dots—many colors, not just one. Attend this image till your mouth waters.

4. Now bite the apple; hear its juicy snap; savor its texture, its flavor. Smell the apple's sweet fragrance.

5. With a knife, slice the apple to see what's inside. As you continue to explore the apple in detail, return occasionally to the larger context; see your hand, feel the soft breeze, be aware of three-dimensionality in form and space.

Multimodal Assimilation

In our culture, we unfortunately tend to repress much sensory experience, making sensory recall difficult. Instead of "taking in the sights" on your next walk, take in the odors; you will be surprised at how little you ordinarily smell. As discussed in Chapter 11, many people also tend to repress tactile sensations. Nonvisual sensory modes are particularly repressed because they are especially related to feelings of pleasure, disgust, and pain. The olfactory pleasure of perfume (and disgust at the smell of spoiled food), the tactual pleasure of sex (and pain from skin abrasion), the kinesthetic pleasure of dancing (and ache of sore muscles), the auditory pleasure of music (and nerve-jangling noise of the city): these feelings that accompany the nonvisual senses are particularly intense. Because we naturally avoid pain and, obeying cultural strictures, also commonly avoid pleasure, we tend to repress much sensory experience. *Sensory experience that is not actively and consciously assimilated is also not readily remembered.*

Full sensory experience, especially in the nonvisual sensory modes that are commonly repressed, is vital to full sensory recall. Schachtel points to the special importance of the nonvisual senses to Proust's *Remembrance of Things Past:*

> In Proust's account, visual sensations are far outnumbered as carriers of . . . memories by those of the lower, more bodily senses, such as the feeling of his own body in a particular posture, the touch of a napkin, the smell and taste of a flavor, the hearing of a sound—noise or melody, not the sound of words. All these sensations are far from conceptual thought, language, or conventional memory schemata.[9]

Proust's sensory remembrances were acute because his conscious sensory experience was vivid and complete; Proust was a delighted student of sense impressions and feelings. In the next exercise (15-8), assimilate an apple as Proust would.

15·8 APPLE III

1. Repeat the previous exercise with a real apple instead of an imaginary one. Savor the apple slowly and pleasurefully, with all of your senses.

2. After eating it, recall the apple in all sensory detail.

Visual Mnemonic Devices

Neisser observes that "the most important advice offered by the many practitioners of memory improvement systems is to develop detailed and articulate schemata into which new material can be fitted." Bruno Furst, for example, teaches people how to remember faces in relation to a chart that visually classifies types of facial structures.[10] Knowledgeable people in every field develop similar associative frameworks to facilitate recall. Such consciously devised cognitive structures, called "mnemonic devices," are most frequently visual in nature.

The ancient Greeks discovered that an imagined building is an excellent mnemonic device. By picturing items to be memorized in various places in the imagined building (for example, the toga that needs mending on the bed, the empty wine jar on the dining table), the memorizer need only tour the building later to recall them. The next exercise (15-9) introduces a similar memory device.

15·9 ONE IS A BUN[11]

Memorize the following rhyming list:

One is a bun,
two is a shoe,
three is a tree,
four is a door,
five is a hive,
six is sticks,
seven is heaven,
eight is a gate,
nine is a line,
ten is a hen.

This is a cognitive structure that will enable you to remember a list of ten objects more accurately and quickly than by rote repetition. Say that you want to memorize the following list:

1. ashtray

2. firewood

3. picture

4. cigarette

5. table

6. matchbook

7. glass

8. lamp

9. wristwatch

10. phonograph.

Form a ludicrous or bizarre visual association between each pair of items in the two lists. Three, for example: a picture perched in a tree. Or nine: a wristwatch hanging on a clothesline. Try it. Mnemonic devices of this sort are well-proven aids to memory.

Emphasis on Exercise

Just as you could not expect to rebuild a weakened muscle overnight, do not expect to revive atrophied inner sensory imagery without repeated exercise. Re-experience relaxed, multimodal assimilation and retrieval every day, experimenting with new subject matter.

References

1. Bartlett, F. *Remembering.* Cambridge University Press.
2. Kubie, L. *Neurotic Distortion of the Creative Process.* Farrar, Straus & Giroux (Noonday Press).
3. McKellar, P. *Imagination and Thinking.* Basic Books.
4. Richardson, A. *Mental Imagery.* Springer.
5. Murphy, G. Quoted by Rugg, H., in *Imagination.* Harper & Row.
6. Perls, F. *Ego, Hunger, and Aggression.* Random House.
7. Huxley, A. *The Art of Seeing.* Harper & Row.
8. Neisser, U. *Cognitive Psychology.* Appleton-Century-Crofts.
9. Schachtel, E. *Metamorphosis.* Basic Books.
10. Furst, B. *Stop Forgetting.* Greenberg.
11. Miller, G., Galanter, E., and Pribram, K. *Plans and the Structure of Behavior.* Holt, Rinehart & Winston.

Recommended Reading

A. R. Luria's *The Mind of a Mnemonist* (Regnery), a classic in the memory literature, describes a rare human being endowed with virtually total recall. Sheila Ostrander and Lynn Schroeder's recent *Superlearning* (Delacorte Press) is a superb experiental book that shows numerous ways to use relaxation and sensory imagination to improve learning and memory. For advanced mnemonic devices, see Bruno Furst's *Stop Forgetting* (Greenberg).

AUTONOMOUS IMAGERY

16

A Primary Source of Visual Imagery

Autonomous imagery is imagery that is not readily susceptible to conscious control. Exemplified by dream imagery, autonomous imagery also takes several other forms, including hypnogogic and hypnopompic imagery (defined below), daydreams, and hallucination. The purpose of this chapter is to point to the productive and creative nature of this primary source of visual imagery. By becoming more aware of dreams and related forms of autonomous imagination, you open a mental door to a primary source of imagery. As suggested in Chapter 4, access to this imagery realm is essential to fully integrated visual thinking.

Hypnogogic Imagery

Hypnogogic imagery is autonomous inner imagery experienced just before falling asleep. Hypnopompic imagery is similar in character but occurs in the drowsy state of coming awake. About one-third of all adults, and considerably more children, experience hypnogogic imagery at least occasionally. Many people, however, have never heard of it. *Knowing what hypnogogic imagery is like, and understanding the conditions that favor it, frequently allows people to experience hypnogogic imagery for the first time, or to experience it more regularly.*

What is hypnogogic imagery like? It is, first of all, stubbornly autonomous. Viewers who attempt to "write the script" of this internal cinema will likely find they cannot. Hypnogogic imagery usually ceases when control is attempted. Efforts to scrutinize this imagery actively—to count the number of windows in the hypnogogic image of a building, for example—will usually dispel the image.

Hypnogogic imagery occurs in all sensory modes, although visual and auditory imagery are most common. In the auditory mode, imagers often hear distinct voices (sometimes their own) or music. In the visual mode, they see a kind of technicolor surrealism. Colors are "more real than real"; grass is "greener than any grass." Detail is also sharply focused and can be acutely distinct.

The content of hypnogogic imagery is extraordinarily varied. Frequently, initial imagery consists of fields or clouds of color, geometric patterns, or life-like pictures that reflect recent periods of prolonged visual stimulation. A person who spends the afternoon weeding the garden, for example, may well experience imagery of plant forms. After hours of driving, the imagery may be of oncoming cars. These are not faithful memory images; memory is infused with fantasy and idealization.

Most hypnogogic imagery consists of dream-like fantasies whose novelty often surprises the imager. "The surprise may be a little one," writes Wilson Van Dusen, "where one can see the associative link. Or it may be like a dream where the message of the inner is not really understandable."[1]

A number of artists—Richard Wagner, Edgar Allan Poe, Max Ernst, and Lewis Carroll, for example—have reported that they have used hypnogogic imagery as source material. In a letter to Peter McKellar, author Ray Bradbury wrote: "Quite often I do discover some preciously good material in the half-awakened, half-

108

slumbery time before real sleep. Quite often I have forced myself completely awake to make notes on ideas thus come upon."[2]

Harold Rugg, in *Imagination*, reports that he typically experiences imagery while he is awakening, either from an afternoon nap or from deep sleep at night. "The hypnopompic state is especially favored. Then, as at no other time, all environmental pressures are off; not only is the body in repose, the mind is relaxed, too. . . . The content of the imagery is succinctly conceptual, has crystal clarity, is marked by sharpness and brevity. Ideas flow in, one after another."[3]

Hypnogogic and hypnopompic imagery can be cultivated. The favorable conditions are essentially those incorporated in exercise 16-1.

16·1 HYPNOGOGIC IMAGERY

---

1. Just before you fall off to sleep, consciously relax your body into a state of deep muscle relaxation (see exercise 6-7). The hypnogogic state favors relaxation almost to the point of falling asleep. When deeply relaxed, your body will be "but vaguely felt, and even more so the contact with the bed sheets and mattress. The spatial position of the body is but poorly localized. Orientation is confused. The perception of time is uncertain."[4]

2. Concurrently, relax your organs of inner speech; silence verbal thinking about events of the day. Allow your mind to become quiet.

3. Also relax your eyes. Paradoxically, you cannot look for a hypnogogic image. Indeed, conscious focusing and scanning (of the sort that normally accompanies perception) will cause hypnogogic imagery to recede: "In order to prolong the phenomenon, a certain 'absence' of voluntary attention is necessary, as in the case of its generation."[4]

4. If you readily fall asleep, "keep your arm in a vertical position, balanced on the elbow, so that it stays up with a minimum of effort. You can slip fairly far into the hypnogogic state this way, getting material, but, as you go further, muscle tonus suddenly de-

creases, your arm falls, and you awaken suddenly."[5]

5. Record your hypnogogic experiences (see exercise 16-3) as soon as possible after they occur; they fade from memory rapidly. Some experimenters have found it possible to record ongoing hypnogogic experience into a tape recorder.

Hypnogogic intervals are characteristically short, but they can be prolonged with practice. What has been said of hypnogogic imagery also holds for hypnopompic imagery, except that reveries experienced upon waking are often influenced by the content of a previous dream.

Daydreams

Entrance to the "hypnogogic cinema" is purchased by relaxed willingness to retreat from sensory input and to entertain inner thoughts and feelings. People who have rigid defenses against impulse life tend to resist such an excursion inward. The opposite is the case of the daydreamer, who commonly prefers inner reality to the outer reality of here-and-now. The daydreamer has ready entrance to the hypnogogic cinema, and indeed frequently calls hypnogogic experience daydreaming.

Not all daydreaming possesses the autonomous spontaneity of hypnogogic imagery, however. The scenarios of many daydreams are enormously predictable; they are "the reverse of a present frustration," writes Frederick Perls. "If broke, we fantasy winning the sweepstakes. If jilted, we wallow in fantasied revenge."[6] The repetitiveness of these compensatory daydreams testifies to their inability to solve problems. *While active visual thinkers direct their fantasies toward expression in reality, compensatory daydreamers escape from reality into fantasy, where they cycle passively and endlessly.*

Not all daydreaming falls under the heading of escape-to-fantasy, however. Many visual thinkers use a form of daydreaming to think productively (16-2). This form of daydreaming possesses the paradoxical characteristic of "purposeful purposelessness."

16•2 LAZIN' DOWN THE RIVER

In a quiet setting, relax and introduce a subject to your mind upon which you would like to think productively. Don't consciously direct your imagination (as you will in the next chapter). Instead, passively and gently follow your thought-stream wherever it may go.

Dreams

As suggested in Chapter 4, making fuller contact with dreams is important to creative visual thinking. Psychologists Stanley Krippner and William Hughes[7] have collected the introspective accounts of a number of well-known thinkers who have experienced creative insights in their dreams. For example, physicist Niels Bohr, in a vivid dream "saw himself on a sun composed of burning gas. Planets whistled as they passed him in their revolutions around the sun, to which they were attached by thin filaments. Suddenly the burning gas cooled and solidified; the sun and planets crumbled away." This dream led Bohr to conceive a model of the atom that had enormous influence on atomic physics. Robert Louis Stevenson "discovered that he could dream complete stories and even go back to them on succeeding nights if the end was unsatisfactory."

Pharmacologist Otto Loewi received a Nobel Prize for a discovery that occurred in a dream after many years of ruminating on a problem concerning the effect of nerves on heart function. When the dream first occurred, Loewi wrote it down and went back to sleep. The next morning he was unable to decipher his notes. That night he had the same dream again, awakened, and this time went directly to his laboratory, where he made an experiment that verified his dream.

Inventor Elias Howe, after years of abortive attempts to develop a sewing machine, dreamed that "he had been captured by savages who dragged him before their king. The king issued a royal ultimatum: if within 24 hours Howe had not produced a machine that would sew, he would die by the spear. Howe failed to meet the deadline and saw the spears slowly raise, then start to descend. Suddenly, Howe forgot his fear as he noticed that the spears all had eye-shaped holes in their tips." Howe awakened, and realizing that the eye of the needle of his sewing machine should be near the point, rushed to his laboratory and soon confirmed the idea.

Another famous inventor, James Watt, invented a process for making lead shot for shotguns in a dream in which "he seemed to be walking through a heavy storm; instead of rain, he was showered with tiny lead pellets." Watts, correctly interpreting his dream "to mean that molten lead, falling through the air, would harden into small spheres," thereupon revolutionized the lead-shot industry.

Krippner and Hughes refer to many other well-known creative thinkers who used dream material in their work: Goethe, William Blake, Edgar Allan Poe, Voltaire, Dante, Shelley, Tolstoy, and Coleridge, among writers; Mozart, Schumann, Saint-Saens, and Vincent d'Indy, among composers; Descartes, Condorcet, and Carden, among mathematicians—and so on. The list of creative dreamers is long and prestigious, and the accounts of creative dreaming hold much in common. We can be quite sure that the phenomenon exists.

How can you experience a creative dream? The initial step toward this goal is to come into better contact with your dreams. Sleep psychologists have shown that everyone has, according to a regular pattern, four to six dreams every night. In a sleep laboratory, periods in which the sleeping subject is dreaming are recorded on an EEG machine; consequently, the dream investigator knows when a subject is dreaming and can wake him or her toward the close of the dream. In this laboratory situation, most dreams are remembered. Under normal sleeping conditions, however, many more dreams are forgotten than remembered. Without bringing scientific instrumentation into your bedroom, how can you remember more of your dreams?

One way to remember more of your dreams is to contrive to sleep fitfully: overeat, for example, and your intermittent wakenings during the night will coincide with the close of more dreams than usual, so that you will likely remember more of them. But this is a drastic measure. A more effective way to make contact with the approximately 90 minutes of dreaming that you experience nightly is to keep a dream diary (16-3).

16•3 DREAM DIARY

1. Just before you fall asleep, repeat the following to yourself several times: Tonight I dream; when I awake I will remember my dreams.

2. "When you first awaken in the morning, lie quietly before jumping out of bed. Let your mind dwell on the first thing that comes up. Do not allow daytime interests to interrupt. Your first waking thoughts may remind you of the contents of your last dream before awakening and allow you, with further practice, to remember more and more details of the dream."[3]

3. Keep a notebook next to your bed in which to keep a diary of your dreams. When you have a particularly vivid dream, also make a sketch of it in your diary. A bedside tape recorder is handy for recording middle-of-the-night dreams: the tape can be translated into the written record of the diary the next morning.

4. Most important, keep the diary daily.

As you accumulate dreams in your dream diary, you will notice that you remember more of your dreams. You will also find new interest in the content of your dreams. Collected together in your diary, repeating or unfolding themes will be especially evident. Although dream interpretation is important to the process of making the unconscious conscious, decoding the metaphorical language of the dream involves knowledge and skills well outside the scope of this book. I can only recommend that you avoid dogmatic interpretations of dream symbols and consult the references at the end of this chapter for a short list of recommended books on the subject of dreams.

Another way to attract dreams to the fore of consciousness is to discuss them with others. Kilton Stewart describes the Senoi tribe of the Malay Peninsula: "Breakfast in the Senoi house is like a dream clinic, with the father and older brothers listening to and analyzing the dreams of all the children." The Senoi not only share their dreams but also use them as a basis for social interaction: a Senoi who dreams of attacking others must "apologize to them, share a delicacy with them, or make them some sort of toy. Thus . . . the tensions expressed in the dream state become the hub of social action in which they are discharged without being destructive."[9] By contrast with the Senoi, our society is close-lipped on the subject of dreams. Another contrast with our society: the Senoi claim that they have not experienced violent crime, war, or serious mental illness for over two centuries.

Senoi children are also taught how to direct their dream experiences. If a child "reports floating dreams . . . he is told he must float somewhere in his next dream and find something of value to his fellows." In other words, the Senoi are taught to dream productively. Exercise 16-4 offers some suggestions on how you can learn to dream productively.

16•4 PRODUCTIVE DREAMING

1. Before you go to sleep at night, review the work you have done on a problem or on a question that has frustrated you. Concentrate for several evenings in a row, if necessary. If you have given the problem enough

pre-sleep attention, you may find upon awakening in the morning that you can remember a dream in which the possible solution appeared.

2. "Try directing your dream thoughts as you might direct your waking consciousness. . . . If your dream seems to be following a negative course, try to reverse it, either in that dream or in a continuation of the dream. If your dream is a positive one, extend it as long as you can and try to derive some use or valuable product from it."[8]

Productive dreaming is strongly favored by long hours of dedicated and conscious effort preceding the dream, as W. I. B. Beveridge points out in *The Art of Scientific Investigation:* "The mind must work consciously on the problem for days in order to get the subconscious mind working on it."[10] The dreamer must also be prepared to unravel the metaphorical language of the dream: Kekulé saw the structure of the benzene ring in the dream symbol of a snake biting its tail; Howe saw the configuration of a sewing machine needle in the symbol of a spear. And not least, the dreamer must be able to move from irrational dream-thinking to logical, disciplined, and reality-oriented thinking in order to verify the insight offered in the dream.

Hallucination

No discussion of autonomous imagery is complete without some mention of hallucination. A hallucination is an inner image experienced as a perception. Abnormal mental states, such as schizophrenia, are characterized by involuntary hallucinations. Hallucinations can also be induced by hypnotic suggestion, by fasting, by lengthy periods of sensory or sleep deprivation, and by hallucinogenic drugs. People working in certain occupations are particularly prone to hallucinations. Truck drivers and jet pilots, for example, report hallucinations that are likely the result of fatigue, monotony, or hypnotic stimuli in the environment.

As dreams can be creative, so can hallucinations; hallucinations are not necessarily delusions to be vigorously avoided. Mystical and visionary experiences frequently have the quality of hallucination; although we commonly deny this kind of experience today, we must nevertheless acknowledge that a visionary such as poet and painter William Blake found much of the beauty that he gave the world in an hallucinatory state of consciousness. And Koestler writes:

> The great physicist Michael Faraday also was a visionary not only in the metaphorical but in the literal sense. He saw the stresses surrounding magnets and electric currents as curves in space, for which he coined the term "lines of force," and which, in his imagination, were as real as if they consisted of solid matter. He visualized the universe patterned by these lines—or rather by narrow tubes through which all forms of "ray-vibrations" or energy radiations are propagated. This vision of curved tubes which "rose up before him like things" proved of almost incredible fertility: it gave birth to the dynamo and the electric motor.[11]

A Society of Half-Thinkers?

Hypnogogic imagery, daydreams, and dreams are given little attention in our outward-oriented society; similarly, the visionary is little tolerated. Our tendency to reject these forms of inner experience may be more sick than sane, however. Kilton Stewart's experiences with the Senoi led him to conclude that

> modern civilization may be sick because people have sloughed off, or failed to develop, half their power to think. Perhaps the most important half. . . . In the West the thinking we do while asleep usually remains on a muddled, childish, or psychotic level because we do not respond to dreams as socially important and include dreaming in the educative process. This social neglect of the side of man's reflective thinking, when the creative process is most free, seems poor education.[9]

We need not give up our power to think logically and rationally in order to explore the thinking that occurs in our dreams. Kekulé, after telling his colleagues how he glimpsed the structure of benzene in a dream, advised, "Let us learn to dream, gentlemen."[11] If we learn to understand, to share, and to act on our dreams, perhaps we will also learn to understand our myths, including our current myth that dreaming is socially useless and that the waking dreams of the visionary are akin to insanity.

Autonomous imagery is a fundamental source of creative thinking. Let us learn to dream.

References

1. Van Dusen, W. "The Natural Depth in Man," in *Person to Person: The Problem of Being Human* (edited by C. Rogers and B. Stevens). Real People Press.
2. Bradbury, R. Quoted by McKellar, P., in *Imagination and Thinking.* Basic Books.
3. Rugg, H. *Imagination.* Harper & Row.
4. Sartre, J. *The Psychology of Imagination.* Philosophical Library.
5. Tart, C. *Altered States of Consciousness.* Wiley.
6. Perls, F., Hefferline, R., and Goodman, P. *Gestalt Therapy.* Julian.
7. Krippner, S., and Hughes, W. "ZZZ Genius at Work ZZZ," *Psychology Today,* June 1970.
8. Krippner, S., and Hughes, W. "Dreams and Human Potential," paper presented to American Association for Humanistic Psychology, 1969.
9. Stewart, K. "Dream Theory in Malaya," in *Altered States of Consciousness* (edited by C. Tart). Wiley.
10. Beveridge, W. *The Art of Scientific Investigation.* Random House (Vintage Books).
11. Koestler, A. *The Act of Creation.* Macmillan.

Recommended Reading

For an excellent review of the role of imagery in creative thinking, along with an unusual series of colored illustrations of autonomous imagery experienced and recorded by the author, see Roger Shepard's chapter "Externalization of Mental Images" in *Visual Thinking, Learning, and Communication,* edited by B. Randhawa and W. Coffman (Academic Press). Jerome L. Singer's *Daydreaming* (Random House) gives a far more thoroughgoing treatment of this subject than is possible here. For an excellent discussion of the metaphorical language of hypnogogic imagery, see Van Dusen's article, cited above. The literature on dreams consists of the theoretical and the experiential. Additional theoretical books on dreams are:

Freud, S. *The Interpretation of Dreams.* Basic Books.
Fromm, E. *The Forgotten Language.* Grove Press.
Hall, C. *The Meaning of Dreams.* McGraw-Hill.
Foulkes, D. *The Psychology of Sleep.* Scribner's.
MacKenzie, N. *Dreams and Dreaming.* Vanguard.
Jung, C. *Man and His Symbols.* Doubleday.

Books that provide a sound experiential approach to dreaming are:

Assagioli, R. *Psychosynthesis.* Hobbs, Dorman.
Perls, F. *Gestalt Therapy Verbatim.* Real People Press.
Martin, P. *Experiment in Depth.* Pantheon.

Patricia Garfield's *Creative Dreaming* (Ballantine) is an excellent guide to using dreams creatively; her *Pathway to Ecstasy* (Holt, Rinehart & Winston) is a candid autobiographical account of her personal exploration into creative dreaming.

DIRECTED FANTASY

17

active fantasizer: a participant, not a spectator. De Mille's games, intended for all ages, open up the freedom of possibility inherent in fantasy and challenge the participants to develop their "imaginative muscles." The following imagination game (17-1) should be played in a quiet environment, free from distraction and interruption; ideally, you should assemble a group of participants and have one person read the game aloud.

17•1 BREATHING

Note to the reader of this game: Ask your listeners to help you obtain a sensitive reading pace by signaling when they are ready to proceed to the next fantasy (a raised hand or nodded head). Pause at each slash (/) for the fantasy to be fully formed by the listeners. *Note to the listeners:* Loosen tight clothing and relax in a comfortable position.

"This game is called *breathing*.

"Let us imagine that we have a goldfish in front of us. Have the fish swim around. / Have the fish swim into your mouth. / Take a deep breath and have the fish go down into your lungs, into your chest. / Have the fish swim around in there. / Let out your breath and have the fish swim out into the room again.

"Now breathe in a lot of tiny goldfish. / Have them swim around in your chest. / Breathe them all out again. /

"Let's see what kind of things you can breathe in and out of your chest. / Breathe in a lot of rose petals. / Breathe them out again. / Breathe in a lot of water. / Have it gurgling in your chest. / Breathe it out again. / Breathe in a lot of dry leaves. / Have them blowing around in your chest. / Breathe them out again. / Breathe in a lot of raindrops. / Have them pattering in your chest. / Breathe them out again. / Breathe in a lot of sand. / Have it blowing around in your chest. / Breathe it out again. / Breathe in a lot of little firecrackers. / Have them all popping in your chest. / Breathe out the smoke and bits of them that are left. / Breathe in a lot of little lions. / Have them all roaring in your chest. / Breathe them out again.

"Breathe in some fire. / Have it burning and crackling in your chest. / Breathe it out again. / Breathe in some logs of wood. / Set fire to them in your chest. / Have them roaring as they burn up. / Breathe out the smoke and ashes.

Controlling Your Imagination

With the important exceptions described in the previous chapter, productive visual thinkers control their inner imagery, manipulate it, transform it, and move it along toward a desired goal. By contrast, worriers are passive victims of negative imaginative ventures which they cannot stop, much less direct. In this chapter, you will learn ways to write the script of the internal cinema of your mind's eye. You need not see clear mental pictures in order to perform the exercises that follow. Be content with whatever clarity of inner imagery you currently possess, and concentrate on developing your ability to control your imagination.

Imagination Games

One way to learn to direct your imagination is by means of "imagination games," such as those suggested by Richard de Mille in his book *Put Your Mother on the Ceiling.*[1] Although written for one person to read to others, de Mille's imagination games are not literary fantasies. Each game casts the listener in the role of an

"Have a big tree in front of you. / Breathe fire on the tree and burn it all up. / Have an old castle in front of you. / Breathe fire on the castle and have it fall down. / Have an ocean in front of you. / Breathe fire on the ocean and dry it up.

"What would you like to breathe in now? / All right. / Now what? / All right. / What would you like to burn up by breathing fire on it? / All right. /

"Be a fish. / Be in the ocean. / Breathe the water of the ocean, in and out. / How do you like that? / Be a bird. / Be high in the air. / Breathe the cold air, in and out. / How do you like that? / Be a camel. / Be on the desert. / Breathe the hot wind of the desert, in and out. / How does that feel? / Be an old-fashioned steam locomotive. / Breathe out steam and smoke all over everything. / How is that? / Be a stone. / Don't breathe. / How do you like that? / Be a boy (girl). / Breathe the air of this room in and out / How do you like that?

"What is the name of this game?"[1]

A more advanced imagination game is Group Fantasy (17-2), which is well worth the trouble of assembling a group. In this game fantasy themes are spontaneously generated and shared by several people.

17·2 GROUP FANTASY

1. Group size: large enough to provide diversity and small enough to allow each member to participate actively — five is ideal.

2. Environment: quiet, secure from intrusion, and preferably dark.

3. Preparation: one good group configuration is a spoke-like pattern with everyone lying on their backs on the floor, heads toward the center. (This pattern combines relaxed posture with easy communication.) Once comfortably situated, spend several minutes relaxing.

4. When relaxed, a self-elected member of the group tells the others of a fantasy he or she is having, describing it as ongoing and in full sensory detail.

5. As in the previous imagination game, each member of the group actively participates in this initial fantasy episode (even joining with appropriate sound effects). Then another member of the group takes over the role of guide, carrying the fantasy further. Then another member comes forth to lead, and so on.

6. The only rule in the game: no criticism is permitted. Fantasy and criticism are not compatible. Any member who cannot contribute constructively should be asked to leave the group.

You can also make up challenging imagination games yourself. "In making up your own games," advises de Mille, "see that the rules of reality are broken as often as they are kept. Water should run uphill. Dogs should meow. Fish should fly."[1] By defying reality, you will clarify the distinction between reality and imagination and strengthen your ability to control imagination.

Overcoming Blocks

Everyone's imagination has built-in "blocks," defenses against thoughts that evoke fear or anxiety. Unwittingly learned and unconsciously used, blocks are difficult to detect and overcome. Self-generated fantasies usually avoid blocked areas of imagination. De Mille, however, purposefully devised his imagination games to lead the fantasizer into areas of experience that are commonly blocked. A daydreamer, for example, would normally avoid many of the imaginative experiences that you just had in the game entitled Breathing. Breathing foreign materials into one's lungs, even in fantasy, can provoke anxiety. But de Mille encourages you to try it. You do; you find that fantasy-things in your lungs are not to be feared; and you overcome a small block. In another of de Mille's 30 imagination games, you are directed into a doctor's office: "Have the doctor say he is going to give you a shot. / Have him say, 'This won't hurt.' / Have him give you the shot. / Have it hurt." A daydreamer would also normally avoid fantasies of this sort. In directed fantasy, de Mille skillfully leads your imagination into other commonly blocked areas. Do you fear your mother? Put her on the ceiling. Do you have a fear of heights? Put yourself on the ceiling. By experiencing these episodes in fantasy, where no real harm can come, you gain confidence and freedom in the use of your imagination. Removing blocks to imagination has the effect of opening doors to a marvelous realm that was there all along.

Group fantasy helps us confront another important and common block to imaginative freedom, the socially induced block. Few adults are able to enjoy the shared imaginative play that is common fare for children. A social situation causes many adults to block their imagination. They fear that their fantasies will appear childish to others, or even somewhat insane. They fear that they will be criticized. Group fantasy provides experience in overcoming blocks to imagination that are essentially social in nature.

While experiencing group fantasy, notice that nothing brings a flight of imagination down more quickly than criticism. An authoritarian figure in the group who refuses to find merit in other people's fantasies quickly smothers the imagination of others in the group. In group fantasy, as in creative social environments generally, permission to be foolish, to be wildly imaginative, to become a child again, must be awarded equally. In an affirmative social climate, every idea has its chance in the sun, and failed ideas are allowed to die quietly.

When working to remove blocks to your imagination, remember that blocks have a function: you have carefully, though unwittingly, constructed them to defend yourself against painful experience. When a block is suddenly removed, you may have to deal with a rush of anxiety-laden experience for which you now have no defense. For this reason, I do not recommend amateur or self-administered "block-busting." You may find that minor blocks will dissolve with patient and gentle effort, but major blocks should be removed only with expert guidance.

Step-by-Step Achievement

What should you do when you come upon a block that you cannot readily overcome? De Mille suggests a technique for overcoming imaginative blocks, which he calls "step-by-step achievement." Suppose that you become anxious when asked to imagine yourself floating up to the ceiling. After reminding you that anything is possible in the world of fantasy (frequently this reminder alone will overcome the block), de Mille would ask you to imagine "a lesser, similar event." In this instance, the first step might be to imagine a balloon floating to the ceiling. Once this easy feat is accomplished in fantasy, progressively more difficult ones are suggested: elevate a book to the ceiling, then a suitcase, a chair, and eventually yourself. "The trick," observes de Mille, "is to find the first, easy step."

Step-by-step achievement can also be used to overcome obstinate imagery. If you have difficulty changing a blue hat into a red one, just change it step by step: simply add a red button to the hat, then two, and so on, until the hat is red. This system can also be used to overcome extremely difficult blocks: psychologist John Wolpe uses step-by-step visualization as a form of psychotherapy.

In Wolpe's therapy, the phobic (fearful) patient is first trained in deep muscle relaxation. (As discussed in Chapter 6, relaxation tends to dispel anxiety.) "Concurrent with this training," writes Wolpe,

we explore the real-life situations that aroused the phobic reaction. We rank the situations in a hierarchy according to how much they disturb the patient. A person with a death phobia, for example, might place human corpses at the top of his list, with funeral processions and dead dogs ranking further down. Then we ask a patient, while he is relaxing, to imagine the weakest item—the one at the bottom of his list. Perhaps this causes some anxiety. Then we ask him to put it temporarily out of his mind and concentrate on relaxing; after about 20 seconds, we ask him to imagine it again. Each time, there is less anxiety, until finally the patient can imagine the scene without anxiety. When the whole list has been treated similarly, it will be found that the real-life situation that created the phobia has also lost its power to produce anxiety.[2]

The fundamental notion underlying hierarchical listing and step-by-step achievement is that *anxiety-provoking imagery should neither be avoided nor confronted head-on.* To avoid a blocked area of imagination is frequently to avoid imagining; to experience the blocked emotion fully and directly usually has the same result. When a block is approached slowly, gently, and through fantasy, imagination can be used to expand itself (17-3).

will not give way to another line of thought. Wolpe suggests two ways to deal with undesired, persistent imagery. The first is called "thought stopping": "The patient puts a recurrent and anxiety-producing train of thought into words. Suddenly the therapist shouts: Stop! We show the patient that his thoughts, in fact, do stop, and eventually he can stop thought on his own." The other method is quite the opposite: "We 'flood' a patient with strong anxiety-eliciting stimuli until his original anxiety is extinguished." Both methods can be self-administered (17-4, 17-5).

17•3 STEP BY STEP

To overcome a block that prevents you from directing your imagination:

1. Describe the block in writing.

2. Decide on an image related to the blocked image but much easier to imagine. Decide on several other related images that are increasingly difficult to imagine. Arrange these in a hierarchy that begins with the easy image and progresses in difficulty toward the blocked image.

3. Relax (see exercise 6-7, Deep Muscle Relaxation).

4. Imagine the easiest (first) image on your list. If you experience anxiety in relation to this image, put it away from your mind's eye, and relax once more. Then try again. Repeat the cycle of imagine-then-relax until you are able to view the first image without difficulty.

5. Go to the next-most-difficult image, and repeat the process until you are able to view this image without anxiety. Continue on up the hierarchy of images: approach the blocked image step by step, gently and patiently.

17•4 STOP!

If you find that you are unable to stop or alter imagery,

1. Verbalize the scenario of the image, and then snap your fingers and order it to "Stop!"

2. Then concentrate on here-and-now sensory reality. (An excellent way to stop insomnia-producing fantasies is to attend to here-and-now body sensations such as progressively relaxing muscles.)

17•5 FLOOD

If you are presented with persistent imagery,

1. Create "more images of the same kind, increase the supply, alleviate the scarcity, and thus reduce the demand."[1]

2. Move closer to the imagery, and then further away; examine it in great detail and from every angle, until the mind is exhausted.

So far, we have been discussing the inhibitive block that prevents entrance to an area of imagination. Now consider the compulsive block: imagination can also be blocked by a persistent train of images that

Self-Directed Fantasy

Once you have the feel of directed fantasy, and a sense of how to encounter inobedient imagery, you can exercise control in many ways. Play imagination games with your environment. (I just ordered my coffee cup to grow wings and fly acrobatic maneuvers overhead.) You can direct your fantasies to change direction rapidly (have a new fantasy every half-minute) and then to stop altogether. Worrying is an excellent reminder to develop your skill at directed fantasy. *Whenever you find yourself worrying, apply a bit of "imaginative muscle" to this negative form of fantasy, and turn it around* (17-6).

17•6 WORRY-IN-REVERSE

Come into control of passive, negative worry and actively direct it into a productive form of fantasy. If you are worrying about a possible failure, envision the pleasurable opposite. If you are worrying that you'll miss a deadline, direct yourself in fantasy to the enjoyable experience of meeting it on time. Then, in a final test of skill, order your fantasies to stop, and direct your attention toward experience of the here-and-now.

"Man becomes great to the extent that he controls his imagination," writes Rolf Alexander, "and impotent to the degree that his imagination controls him."[3]

Imaginative Viewpoints

As you learn to control your imagination, also begin to experiment with imaginative viewpoints. From one viewpoint, for example, you can move yourself through space, perhaps by ice-skating or flying. From another, you can maintain your position and view an object as it moves around you, as in the next fantasy (17-7).

17•7 OBJECT-IN-SPACE

With your eyes closed, imagine that you are standing motionless on a desert plain. It is dark. /

In the distance, you hear a guitar. / You can see the guitar, far-off / magically suspended in mid-air / lighted by a theatrical spotlight.

Now the guitar floats, slowly toward you, positioned vertically, face forward. / The music becomes more distinct as the guitar comes closer to you. / Stop the guitar three feet away from you so that you see its distinctive shape / its rich wood grain and sparkling finish. / Examine its details of mother-of-pearl / black ebony / ivory and brass / taut strings.

Now slowly rotate the guitar on its vertical axis so that you can view its side / then its back. / For the next few moments, also rotate the guitar on its horizontal axis / again savoring its beautiful details and finish.

Now move the guitar into the distance on the desert plain. / Allow the music to fade / and dim the spotlight.

In your imagination, you can effortlessly move *anything* through space.

As you can position your viewpoint in space, you can also position it in time. In Chapter 15, you moved your viewpoint into the past; in exercise 8-1, Feeling the Actual, you situated your awareness in the immediate present; in Chapter 19, you will move it into the future.

Another valuable control is to move from the distant viewpoint of observer (as in the previous fantasy, for example) into the merged viewpoint in which you identify with your imaginings (17-8).

17•8 MERGED

Close your eyes / and imagine that you are a beautiful red rose bud / on a rose bush / in a lovely country garden. / You are not observing a rose bud "out there." / You *are* the rose bud / and you are ready to bloom. / You can feel the warm sun on your outer petals now / the cool, fragrant moisture within. / Each of your petals is growing / delicately pressing outward. / Feel a marvelous sensation of relaxation pour through your rose-body / as you open your petals luxuriantly / to the warmth of the morning sun. / Smell your delicate fragrance / feel your velvety red texture. / A cool drop of morning dew is forming at your very center / sparkling like a watery diamond. / You *are* a perfect rose / on a rose bush / in a country garden / basking in the morning sun.

The merged viewpoint is especially effective for obtaining insights in problem-solving. For example, merged, you can imagine what it would feel like to be some aspect of your problem. What would it feel like, for example, to be a stylus moving through the groove of a record? What would you say to the bottle if you were the cap? The observer viewpoint has the opposite virtue of permitting you to stand back from the problem to obtain both physical and psychological distance. How important does your problem seem, for example, when viewed from the moon? The ability to move your imaginative viewpoint in space and time is truly one of the most valuable powers of directed imagination.

References

1. De Mille, R. *Put Your Mother on the Ceiling.* Walker.
2. Wolpe, J. "For Phobia: A Hair of the Hound," *Psychology Today,* June 1969.
3. Alexander, R. *The Mind in Healing.* Dutton.

Recommended Reading

Roberto Assagioli's *Psychosynthesis* (Hobbs, Dorman) suggests a number of directed fantasy themes that are psychotherapeutic, even spiritual, in intent. Joseph Wolpe, in *The Practice of Behavior Therapy* (Pergamon Press), presents, by means of a number of fascinating case studies, the use of fantasy and deep relaxation to desensitize phobia. The student of Eastern religious practices will find references in this literature to visual meditation fantasies directed toward the goal of spiritual realization. A. R. Orage's *Psychological Exercises and Essays* (Janus Press) contains many brief fantasy exercises similar to those in this chapter.

STRUCTURES AND ABSTRACTIONS

18

Beneath the Surface

Implicit in every act of perception is knowledge of what lies beneath the visible surface. I look at an orange: I perceive that it is not hollow, like a child's rubber ball, but that its skin contains fruit. I look at my hand: I perceive skin, hair, and fingernails, but my imagination contributes more, for I also perceive that my skin envelops a living structure of bones, tendons, blood vessels, nerves, and muscles. I look at my desk: I perceive that it has a back side and that inside its drawers are papers, pencils, and other possessions. I do not experience the orange, my hand, and my desk as mysterious, opaque balloons. Memory joins with sensation of outer surfaces, and I perceive objects that have insides as well as outsides.

In the exercise that follows (18-1), do what the man in Cobean's cartoon (Figure 18-1) is doing: explicitly peel away surfaces to imagine what is underneath. But go further: remove all outer layers to visualize internal structure. *The ability to comprehend structure is essential to visual thinking.* Thinking that occurs only in relation to visible surfaces is superficial; it is only skin-deep.

18·1 X-RAY VISION

1. Overlay a magazine photograph with tracing paper, and draw (to the best of your knowledge) what lies beneath the visible surface. As you seek the internal physical structure of the object (using several colors, if necessary) also comprehend it. Ask yourself, for example: "What does this bone do?" or "What is this connection for?"

2. Perform the same operation on an actual object; in your mind's eye, visualize the internal structure of your hand, for example. Try doing it with your eyes open: juxtapose a mind's-eye image of inner structure upon a perceptual image of external surfaces.

In Section V, you will explore the penetrating and dissecting capabilities of your visual imagination further, in the context of idea-sketching. The visual languages of perspective and orthographic projection are especially well suited to exploring internal structure.

Manipulating Structure

Visual thinking is actively operating upon structure, not only to see what is inside, but also to manipulate the component parts of structure in relation to each other. Adjusting proportion relationships (18-2) is one kind of manipulation of structure in which only sizes are changed.

18·2 RATIO

Imagine that you have a magic powder called Ratio, which makes things larger or smaller, as you wish. In your mind's eye, sprinkle Ratio on a friend, shrinking him or her to half-size. Ratio can also be applied selectively, increasing the size of one part, and decreasing the size of another. Experiment with your current setting: for example, make the ceiling higher, the windows larger, the furniture more delicate in scale, and so on.

Figure 18-1.
Drawn by Cobean; copyright © 1950 The New Yorker Magazine, Inc.

A somewhat more difficult mental manipulation involves keeping proportion the same while changing the position of the structure's component parts. An automotive engineer who looks under the hood of an automobile not only "sees through" to internal structure but also visualizes parts moving in relation to each other: pistons going up and down, the crankshaft going around, and so on. The same kind of imaginative manipulation of structure is important in many activities: the doctor visualizes a moving knee joint, not a fixed one; the physicist sees particles in motion; the businessman imagines cash flow. Exercise 18-3, a two-dimensional exercise taken from a psychological test, gives excellent training in visualizing changes in structural position.

18•3 CHANGING PLACES

You will need the assistance of another person and a piece of white paper and five small shapes of colored paper (a yellow triangle, a blue circle, a black square, a purple cross, and a red diamond).

1. Have the person assisting randomly place one colored shape in each corner of the white square, and the remaining shape in the center.

2. Take one minute to commit your arrangement of colored shapes to memory. Then close your eyes.

3. The person who is assisting this exercise should change the position of two shapes, announcing, for example: "I am switching the black square and blue circle." After a five-second pause, another switch should be made and described verbally.

4. Without looking up, open your eyes and draw your mental image of the current arrangement of colored shapes. Then check your drawing with the actual arrangement.

5. If the exercise is too difficult, reduce the number of shapes to three and the number of switches to one, and increase difficulty gradually. With repeated practice, this mental operation becomes easier. When you master this exercise, you'll be able to track manipulations of the five-shape pattern indefinitely in your mind's eye.

Do not be surprised if you find that mental manipulations of structure are difficult to make. The sculptor who physically changes the proportion of a clay figure performs work. The visual thinker who does the same operation mentally also performs work. Rearranging your living room furniture takes muscles. Similarly, moving component parts in your mind's eye requires "imaginative muscles"; indeed, you may actually experience sensations in your muscles (kinesthetic imagery) while mentally manipulating structure. Whether performed physically or mentally, moving things around requires active effort. It is always easier to leave things as they are.

Complex Mental Manipulations

Mental tasks become more arduous with every step toward complexity. Three-dimensional tick-tack-toe is more difficult than the two-dimensional child's game. Chess, with 64 positions, requires far more mental effort to master than the previous exercise, which has only five. Let us briefly consider the complex game of chess.

People who program computers to play chess are far from being able to develop a chess-playing computer that can rival a human chess grand master. To do this, they realize, would require that they penetrate to the very core of human thinking. Psychologists, inspired by the same notion, have studied the mental processes of chess grand masters. What they have learned tells us much about visual thinking.

You perhaps have heard about the incredible long-term memories of chess masters; some are apparently able to recall every move of every game they have played. Dutch psychologist de Groot has found that the short-term memories of chess masters are equally acute. He showed an unfamiliar chess situation to chess players of various strengths for five seconds; expert and average players made many errors in reconstructing the chessboard from memory, while players

such as the former world champion Max Euwe recalled each situation perfectly.

Binet and others have found that chess masters rarely see a realistic and detailed memory image of the chessboard, however. Instead, they commonly see a gestalt-like image made up of strategic groupings. Their inner imagery is pattern-like (see Chapter 10). Pillsbury reported a "sort of formless vision of the positions"; Alekhine said he visualized the pieces as "lines of force."[1] Chess is extremely complex: each move is made in relation to an incredible number of alternative moves and countermoves. Expert chess players cannot allow their thinking to be distracted by irrelevant details; they think in relation to an abstract sensory image, not a concrete one.

Much more has been written about thinking in chess, but these brief observations are sufficient to introduce an important characteristic of thinking by visual images. Complex thinking operations often require imagery that is abstract and pattern-like. Which is not to say that abstract imagery is more important than concrete; rather, *abstract and concrete imagery are complementary. The flexible visual thinker moves readily back and forth between the two.*

So far, you have dwelt on the experience of concrete mental pictures. Now let us build a case for abstraction, for imagery that embodies the essence of structure without its sensuous qualities of detail. "The test of an image," writes E. H. Gombrich, "is not its likeliness but its efficacy within a context of action."[2] Anton Ehrenzweig alludes to a similar point when he observes that "at a certain point which has to do with the awakening of creativity, the student has to learn to turn about resolutely and blur his conscious visualization in order to bring into action his deeper faculties."[3]

Abstract Visual Thinking

Napoleon is said to have held that those who think only in relation to concrete mental pictures are unfit to command. He reasoned that the commander who enters into battle with a detailed image of his battle plan fixed in his mind finds that image difficult to modify quickly to accommodate sudden and unpredictable changes on the battlefield. Sir Frederick Bartlett concurs: "Too great individuality of past reference may be very nearly as embarrassing as no individuality of past reference at all."[4] Peter McKellar suggests why this is frequently so, pointing to the tendency of concrete imagery to solidify thinking prematurely: "Intelligent thought is inhibitory in the sense that it has stopped short of greater concretization, and been halted at a point which still allows a wide choice from a range of alternative responses."[5]

What is the nature of an abstract inner image? According to Rudolf Arnheim it is, first of all, "often faint to the extent of being barely observable"—indeed, so faint that it "may not be readily noticed by persons unaccustomed to the awkward business of self-observation."[6] Abstract visual imagery especially eludes the introspective observer who equates mental pictures with photographic realism. The pattern-seeking exercises in Chapter 10, in which the gestalt of the image is prominent and details are subordinated, offer a clue to the nature of abstract inner imagery. The schematic idea sketches in Chapter 21 offer another clue; abstract inner images are similar to schematic drawings. Topology, a branch of geometry that deals with spatial equivalents, suggests another analogy. Topologically, the earth and an apple are the same; so are a doughnut and a lifesaver: the generic sphere and torus embody only the abstract essence of structure, and not its specific attributes.

These examples, however, do not afford an experience in *thinking* in relation to abstract inner images. *The distinction between abstract images used as symbols for communication and abstract images used as vehicles for thinking is a vital one.* I will attempt to illustrate this difference by going directly to the use of stereotyped abstract images in language, and from there, to the formation of abstract images in productive thinking.

The Visual Stereotype

Abstract visual images are closely allied to words. Take the word "flower," for example. As a sound issuing from your mouth, or as a series of ink marks on this page, the word "flower" is meaningless. The meaning of the word "flower" is not in the word but in the image to which the word refers. This image, at least initially, is simple and stereotyped; it is an image that captures the essential similarity of all flowers. Stereotyped use of language can diminish sensory experience: out of laziness and habit, many of us see green blobs on a stick when we hear the word "tree." Little effort is required to experience stereotyped imagery of this sort. Ancient languages, such as Chinese, contain many such abstract, pictographic symbols.

The next exercise (18-4) is intended to help you seek abstract inner imagery associated with language; from this experience you may more easily recognize abstract imagery that is a vehicle of thinking.

18•4 ABSTRACT WORD-IMAGES

Take the following abstract words one by one and experience the abstract inner image elicited by the word. Make an abstract sketch of each image.

1. Nouns: chair / tree / house / car / animal.

2. Verbs: thrust / shut / penetrate / collapse / swing.

3. Adjectives: turbulent / sharp / voluptuous / decayed / lively.

Did you find the exercise difficult to perform? Did you, for example, find that a specific chair comes to your mind's eye when the word "chair" is mentioned, but not an abstract visual image of universal "chairness"? Did you experience even more difficulty with the verbs and adjectives? "Shut" can refer to many things: shut the door, shut your eyes, shut off the water. "Turbulent" what? turbulent air, turbulent water, turbulent times?

One way to perform exercise 18-4 is to remember a previously seen abstract graphic representation of the word. A kindergartener could interpret the nouns this way. The verbs and adjectives are more difficult: let me see, an arrow for "thrust," but what for "shut"? Another way is to rummage about in memory for the concrete instances that the word suggests: the word "bird," for example, elicits images of canaries, eagles, ostriches, flapping wings and pecking beaks, bird songs, geese migrating in formation, a hawk soaring effortlessly, a pelican plummeting into the ocean after a fish, and so on. With some effort I can abstract from these concrete instances a visual essence that has a universal quality of "birdness." The latter course, more likely to produce an original result, characterizes one kind of search for abstract imagery in art.

Seeking abstract visual imagery to represent existing concepts is one thing; using it to *develop new ideas* is quite another. The visual thinker who uses abstract inner imagery to develop an idea uses it dynamically, much as you manipulated the colored shapes in the experience earlier in this chapter. Such imagery may be an abstract three-dimensional structure; it may also be a vague "pattern of forces" such as reported by the chess player. *Although frequently vague and elusive to the mind's eye, abstract inner imagery is a primary vehicle for the act of creation.* We turn now to this crucial mental operation.

Hidden Likenesses

Jacob Bronowski writes: "The discoveries of science [and] art are explorations—more, are explosions of a hidden likeness. The discoverer or artist presents in them two aspects of nature and fuses them into one. This is the act of creation, in which an original thought is born, and it is the same act in original science and original art."[7]

Once discovered, hidden likenesses can be described in many ways. In the verbal arts, a hidden likeness is encoded in a simile, analogy, or metaphor. Similes and analogies point to likenesses explicitly (for example: "The Renaissance was like the opening of a flower"); metaphors do so implicitly ("The Renaissance blossomed").

On the usual conscious level of language, of course, there is no likeness between flowers and the Renaissance. The hidden likeness is on a deeper level, beyond words, where sensory and emotional memories associated with the two words overlap. The correspondence between a flower blossoming and the rebirth of human spirit that occurred in the Renaissance enriches the meaning of language in a sensory way. But the correspondence itself is initially seen abstractly. The inventor of the metaphor suddenly sees an abstract connection between the two concepts—a hidden likeness barely glimpsed in a bridging abstract image. The power of the metaphor is usually proportional to the dissimilarity of the ideas that the metaphor links.

As you will remember from Chapter 15, access to vivid sensory and affective memories, to that portion of memory containing material for the discovery of vivid and illuminating hidden likenesses, often requires the relaxation of conscious control. The poet will confirm this: the creative metaphors with which the poet takes language by surprise are rarely the product of conscious and logical effort. Thus the productive abstract image is vague and even subliminal because it occurs subconsciously. As Lawrence Kubie observes:

> Preconscious processes are not circumscribed by the more pedestrian and literal restrictions of conscious language. . . . Preconscious processes make free use of analogy and allegory, superimposing dissimilar ingredients into new perceptual and conceptual patterns, thus reshuffling experience to achieve that fantastic degree of condensation without which creativity in any field of activity would be impossible. In the preconscious use of imagery and allegory many experiences are condensed into a single hieroglyph. [For "hieroglyph" read "abstract inner image."][8]

Is it possible, then, to direct your conscious attention to an abstract inner image that is not stereotyped? Probably not until you have grappled for a while with the elements of a problem that is meaningful to you. Also probably not until your usual conscious mode of thought is relaxed, or put off guard. Exercise 8-5, Visual Similes, will give you practice in finding hidden likenesses in the context of seeing. Decoding your own dream imagery provides a related but reverse experience. The language of dreams is metaphorical, but a dream presents only one element of the metaphor. You discover the other when you consciously interpret the meaning of the dream. Another way to obtain experience in forging abstract, metaphorical resemblances is provided by a thinking strategy called Synectics.

Synectics

Synectics is a problem-solving technique that uses metaphors and analogies to generate creative ideas. I cannot do justice to the rationale of Synectics here; the method is treated fully in *Synectics* by W. J. J. Gordon[9] and *The Practice of Creativity* by George M. Prince.[10] Nevertheless, follow along on a "Synectics excursion" (18-5).

18•5 SYNECTICS EXCURSION

Imagine that you are a silent member of a small group as it explores metaphors related to a problem. As you do, be aware of inner imagery, concrete *and abstract,* that is elicited by the dialogue.

A Synectics group was asked to invent a new kind of roof. Analysis of the problem indicated that there might be an economic advantage to a roof that is white in summer and black in winter. The white roof would reflect the sun's rays in summer so that the cost of air conditioning could be reduced. The black roof would absorb heat in winter so that the cost of heating could be minimized. The following is an excerpt from the session on this problem:

A: What in nature changes color?

B: A weasel—white in winter, brown in summer; camouflage.

C: Yes, but a weasel has to lose his white hair in summer so that the brown hair can grow in. Can't be ripping off roofs twice a year.

E: Not only that. It's not voluntary and the weasel only changes color twice a year. I think our roof should change color with the heat of the sun. There are hot days in the spring and fall—and cold ones too.

B: Okay. How about a chameleon?

D: That is a better example because he can change back and forth without losing any skin or hair. He doesn't lose anything.

E: How does the chameleon do it?

A: A flounder must do it the same way.

E: Do what?

A: Hell! A flounder turns white if he lies on white sand and then he turns dark if he lands on black sand—mud.

D: By God, you're right: I've seen it happen! But how does he do it?

B: Chromatophores. I'm not sure whether it's voluntary or nonvoluntary. . . . Wait a minute; it's a little of each.

D: How does he do it? I still don't plug in.

B: Do you want an essay?

E: Sure. Fire away, professor.

B: Well, I'll give you an essay, I think. In a flounder the color changes from dark to light and light to dark. I shouldn't say "color" because although a bit of brown and yellow comes out, the flounder doesn't have any blue or red in his register. Anyway, this changing is partly voluntary and partly nonvoluntary, where a reflex action automatically adapts to the surrounding conditions. This is how the switching works: in the deepest layer of the cutis are black-pigmented chromatophores. When these are pushed toward the epidermal surface the flounder is covered with black spots so that he looks black—like an impressionistic painting where a whole bunch of little dabs of paint give the appearance of total covering. Only when you get up close to a Seurat can you see the little atomistic dabs. When the black pigment withdraws to the bottom of the chromatophores, then the flounder appears light colored. Do you all want to hear about the Malpighian cell layer and the guanine? Nothing would give me greater pleasure than to . . .

C: You know, I've got a hell of an idea. Let's flip the flounder analogy over onto the roof problem. Let's say we make up a roofing material that's black, except buried in the black stuff are little white plastic balls. When the sun comes out and the roof gets hot the little white balls expand according to Boyle's law. They pop through the black roofing vehicle. Now the roof is white, just like the flounder, only with reverse English. Is it the black-pigmented part of the chromatophores that come to the surface of the flounder's skin? Okay. In our roof it will be the white-pigmented plastic balls that come to the surface when the roof gets hot. There are many ways to think about this."[9]

As the Synectics excursion continues, the participants seek additional analogies between their problem and such diverse realms as biology, art, and history. If they are fortunate, a metaphor connects: *in a flash of excitement, a "hidden likeness" is found that illuminates a solution to the problem. The connecting image is abstract, but it is not stereotyped.* As this abstract inner image rises into conscious awareness, it is recognized as the essential structure of a new idea.

Henri Bergson expresses only half of the truth when he says, "To work intellectually consists in conducting a single representation through different levels of consciousness in a direction which goes from the abstract to concrete, from schema to image."[11] Arnheim gives us the other half: "The mind is just as much in need of the reverse operation. In active thinking, notably in that of the artist or the scientist, wisdom progresses constantly by moving from the more particular to the more general."[6]

The visual thinker who moves along the path from the abstract to the concrete is exercising one of the most powerful thinking strategies. In abstraction, the thinker can readily *restructure* a concept, or even *transform* it. The resulting repatterned abstraction can then be concretized into a form that can be tested in the cold light of reality. The strategy works equally well in the opposite direction, as when a fresh look at a concrete image suggests a new abstraction. When abstract and concrete ideas are expressed in graphic form, the abstract-to-concrete thinking strategy becomes visible: see Chapter 21 for idea sketches and further discussion that will help you master this valuable thinking tool.

References

1. Hearst, E. "Psychology Across the Chessboard," *Psychology Today,* June 1967.
2. Gombrich, E. *Art and Illusion.* Princeton University Press.
3. Ehrenzweig, A. "Conscious Planning and Unconscious Scanning," in *Education of Vision* (edited by G. Kepes). Braziller.
4. Bartlett, F. *Remembering.* Cambridge University Press.
5. McKellar, P. *Imagination and Thinking.* Basic Books.
6. Arnheim, R. *Visual Thinking.* University of California Press.
7. Bronowski, J. *Science and Human Values.* Harper & Row.
8. Kubie, L. *Neurotic Distortion of the Creative Process.* Farrar, Straus & Giroux (Noonday Press).
9. Gordon, W. *Synectics.* Harper & Row.
10. Prince, G. *The Practice of Creativity.* Harper & Row (Harper Torchbooks).
11. Bergson, H. Quoted by Sartre, J., in *The Psychology of Imagination.* Philosophical Library.

Recommended Reading

Samples, B. *The Metaphoric Mind.* Addison-Wesley.

FORESIGHT AND INSIGHT

19

Your Distance Sense

Taste, touch, and kinesthesia enable you to sense only what is within reach of your body. Hearing and smell extend your sensory domain several miles at most. Yet you can see the stars: vision is by far our most effective "distance sense." The next exercise (19-1) challenges you to use your visual sense predictively—that is, to look ahead in time as well as space.

19•1 LOOK AHEAD

Starting with the entrance arrow on the left of the maze in Figure 19-1, draw a pencil line to the exit arrow in the lower right. If you enter a blind alley, return to where you entered it and continue on. The pencil line may not, of course, cross any "wall" of the maze. You may wish to overlay a piece of tracing paper.

Figure 19-1.
From V. Koziakin, *Mazes* (Grosset & Dunlap).

Two Kinds of Foresight

To foresee is to have a mental picture of something to be, to imaginatively envision the future. I am not describing a rare, occult power. *Virtually everyone exercises foresight:* worry and anxiety could not exist without it. Few people know how to use foresight productively, however. Foresight is extremely powerful when used to envision *future goals.* And it is also an invaluable faculty when used to envision alternative *future consequences* of present plans. In this chapter, you will experience these two kinds of foresight. Subsequently you will see how foresight and creative insights are related.

Old-Fashioned "Vision"

Leaders in every field of endeavor are often individuals with a powerful vision of the future. The terms "vision" and "visionnary" may seem rather out-of-date; nevertheless, history has been written by men and women with impossible dreams, figures so effective that they may seem to some to possess mysterious powers. Perhaps the only mystery is how visionaries learn to use their foresight creatively, while most of us, influenced by an education that dwells mainly on the past, never learn the immense power of envisioning future goals.

Maxwell Maltz, in his popular book *Psycho-cybernetics*, develops a sound psychological case for the effect of sustained and affirmative foresight upon behavior. The basic premise of psycho-cybernetics is that the human psyche is goal-seeking and that imagination is the psyche's "steersman" ("cybernetics" comes from a Greek word meaning steersman). Writes Maltz: "We act, or fail to act, not because of will as is commonly believed, but because of imagination."[1]

The technique of psycho-cybernetics is disarmingly simple. First, decide on a goal: let us say, for example, that you want to be more skillful in drawing. Second,

regularly imagine yourself, as vividly as possible, having fully attained your goal. In this example, imagine yourself drawing skillfully. In a recent psychology experiment a number of basketball players were divided into two groups of equal skill. The first group was asked to spend a period of time each day on the court, improving their skill shooting baskets. The second group was asked to spend the same period of time off the court, *vividly imagining themselves shooting baskets with breathtaking accuracy.* Afterwards, both groups were tested to determine which one had actually improved the most. The team who had practiced in their imagination was significantly more improved.

The theoretical foundation and experimental validation of psycho-cybernetics is documented in Maltz's book and elsewhere in psychological literature. It does work, if you practice it faithfully (Maltz suggests that you reserve judgment for a minimum period of 21 days of sustained practice). In exercise 19-2, try the technique first on a minor goal. As you get a feel for it, increase the magnitude of the goal in the direction of an impossible dream.

19·2 ENVISION A GOAL

---

1. Envision a goal: perhaps a desired new behavior or a resolution to a current problem.

2. Set aside 10 minutes just after waking and 10 minutes just before retiring to experience this goal in your imagination. Relax comfortably in a seated position. Close your eyes.

3. Picture yourself having attained your goal *here-and-now.* In your imagination, experience the desired behavior as attained, the unresolved problem as resolved. For example, if you wish to lose 10 pounds, envision being 10 pounds lighter *now.*

4. "The important thing is to make these pictures as *vivid* and as *detailed* as possible. . . . The way to do this is to pay attention to small

details, sights, sounds, objects, in your imagined environment. . . . If the imagination is vivid enough and detailed enough, your imagination practice is equivalent to an actual experience, in so far as your nervous system is concerned."[1]

5. Along with an intense sense of being there, imagine the *positive feelings* evoked by the realized accomplishment.

6. That's all there is to it. Don't dwell on the goal, don't make a time-tabled plan to achieve it, and don't constantly evaluate how you are progressing. Outside your daily experience of goal-setting, stay in the "here-and-now."

You may want to return to Chapter 15, Visual Recall, and Chapter 17, Directed Fantasy, for additional hints relevant to this exercise. The exercises in the chapter on visual recall are directed toward obtaining *clarity* of imagery; the directed fantasy exercises, toward obtaining *control*.

Alternative Futures

Henry Ford clearly foresaw a need for mass-produced automobiles. Ford and his colleagues were markedly bereft of foresight, however, regarding the future consequences of a society on wheels. With 20/20 hindsight, you and I can make a lengthy list of these positive and negative consequences. The question is: Could these consequences have been foreseen?

Living with unforeseen consequences such as smog and traffic congestion, we are coerced into awareness of a need for foresight about the possible effects of present plans. What will be the ecological effect of new forms of air transportation? Is current education preparing our children for the future? How will the computer and automation influence patterns of employment? In an era of rapid change, we badly need to foresee the future in order to prepare for it.

Awareness of the need for foresight is not the same as the ability to foresee, however. *Envisioning future consequences is an art form that badly needs to be educated.* Richard de Mille suggests a way that history could be taught to educate this kind of foresight. Instead of merely teaching that Matthew Perry's visits to Japan in the 1850's brought Japan into the modern world, de Mille suggests that the teacher might ask, "What would have happened if, in 1808, Perry had joined the Army instead of the Navy?" At this point, students could foresee many alternative possibilities: "Japan might still be feudal today. Or Japan might now be part of the Soviet Union. Since there would be no Pearl Harbor attack, the United States would have stayed out of World War II."[2] And so on. De Mille's suggestion has many advantages, not the least of which is the way it orients the student creatively to the future as well as to the past.

One can rarely foresee the actual future consequences of present plans, of course. Most plans are realized in a context that contains many variables and even surprises over which the foreseer has little or no control. One can foresee "alternative futures," however, a bracketed set of possibilities within which the future will likely occur.

"Futurists," using techniques which are too varied and complex to be described here, are currently writing "scenarios" or alternative histories of the future (of the world to the year 2080, for example). The next exercise (19-3) is intended to give you a mini-experience in being your own futurist.

19•3 ENVISION CONSEQUENCES

1. Conceive a decision that would likely influence your personal future. Commit the decision to writing.

2. Also write down approximately a dozen conditions that will likely influence the outcome of your decision (health and economic factors, for example).

3. By varying these dozen conditions (say your health becomes better, stays the same, or deteriorates) and by combining them into different patterns, write several "scenarios" describing the possible year-by-year consequences of your decision over the next five years.

4. Has this exercise in foresight influenced your initial decision?

Insight: Like Lightning

"One hears—one does not seek; one takes—one does not ask who gives: a thought flashes out like lightning inevitably without hesitation—I have never had any choice about it." *Like lightning.* Nietzsche's simile captures the essence of insight. We have discussed foresight. Now let us examine three of the conditions that foster insight: preparation, incubation, and the defocused state of consciousness that usually accompanies an "intuitive leap."

Intensive *preparation* almost always precedes the sudden flash of vision known as insight. "Discovery," writes Jerome Bruner, "favors the well-prepared mind."[3] The insights obtained in the creative dreams discussed in Chapter 16 were obtained by dreamers who had long struggled with their problems. *Sudden illumination is rarely easily obtained: though brief itself, insight usually follows a prolonged contest with the elements of a problem.*

Insight also follows *incubation*, a period of time during which thinking proceeds subconsciously. Most creative people know the value of putting their work aside and turning to something else for awhile. Relaxing, taking a walk, or "sleeping on it" can be extremely productive, although it may appear to be mere laziness to those who identify thinking solely with hard work. Incubation (19-4) is preceded by hard work: the baker who rests while the oven bakes the bread must first mix the ingredients and knead the dough.

19•4 INCUBATE

After intensively working on a problem, put it aside for a while and let your subconscious do the work. Attend another task. Better yet, relax. Even better, sleep on it.

Kelly observes that "it is curious that modern psychology cannot account wholly for flashes of insight of any kind, sacred or secular."[4] Insight, in its unpredictability and infrequency, is clearly not the sort of mental phenomenon that can be readily observed and replicated in a psychology lab. But even more important, insight rarely occurs in the "normal" state of waking consciousness. As Tyrell suggests, "Those creations of the human mind which have borne pre-eminently the stamp of originality and greatness have not come from the region of consciousness. They have come from beyond consciousness, knocking at its door for admittance."[5] Modern psychology has learned little about insight largely because it has been concerned primarily with waking consciousness, virtually ignoring that this "is but one special type of consciousness, while all about it, parted from it by the filmiest of screens, there lie potential forms of consciousness entirely different."[6]

Creative insights are rarely found by straining for them consciously. They come when normal waking vigil is relaxed—in an off-guard reverie, in a daydream, in a dream. Most of all, they come unexpectedly, not as the foreordained product of step-by-step reasoning but as the uncalculated result of an "intuitive leap" (19-5).

What can be said, then, about the experience of insight itself? Rercad what was said in the previous chapter about the discovery of hidden likenesses. Then understand that *insight comes most readily to those who are open to the possibiilty of intuitive leaps.* "My advice," writes Bruner, "in the midst of the seriousness, is to keep an eye out for the tinker shuffle, the flying of kites, and kindred sources of surprised amusement."[3]

19·5 LEAPING

Guesses, hunches, or intuitions are skilled leaps over territory usually crossed by short steps. As with any skill, intuitive ability improves with informed experience. Practice making guesses before information is dutifully worked over, and listening for answers that emerge from subconscious levels of thinking. Gradually you will learn to trust hunches that arrive unexpectedly by avenues that do not appear on the logician's map of the intellect.

Foresight and Insight Are Related

Insight follows preparation and incubation. But preparation presumes a problem. How did the thinker find a problem worth thinking on? Someone, perhaps not the actual problem-solver, *foresaw* the need for an answer and set a goal. Insight cannot occur without foresight. *Insight, sudden and brief, is a mental explosion of energy provided by foresight, long and sustained.*

Indeed, foresight provides goal-tension that pervades every mental operation of problem-solving. Foresight influences what is seen and not seen in a problem situation, and which potential solutions are valued and which are rejected. In short, foresight, however vague, pulls the thinker in the direction of a solution. When problems are assigned, as they commonly are in school and in industry, the impetus provided by foresight is usually diminished and distorted. Specialization that separates problem finders from problem solvers (say, market researchers from designers) can severely hinder the

energy and integrity that foresight imparts to insightful problem-solving.

Finally, foresight is related to insight after-the-fact. Is the insight valid? Envisioning "alternative futures" for an insight is an important way to test its merit.

Limits to Imagining

At the end of the last section, I described several important limitations of seeing as a mode of visual thinking. What are the limitations of imagining? A major deficiency is imagination's ephemeral nature. The internal cinema rarely plays twice: inner imagery is always reconstructed differently for each re-viewing, and sometimes (as with dreams) cannot be reviewed at all. Sir Frederick Bartlett describes another pitfall:

> When once the image method has been adopted and practiced . . . it tends itself to become a habit. . . . A typical visualizer, for example, often seems to have a great wealth of images . . . so that he is tempted to stop and describe them—often to his own and others' aesthetic enjoyment—instead of concentrating upon the problem that they are there to help him to solve. Before long his images themselves get into ruts . . . and lose their touch with [the environmental reality] without which the method of images would originally never have developed at all.[7]

And Miller, Galanter, and Pribram add: "Unless you can use your image to do something, you are like a man who collects maps but never takes a trip."[8]

In this section, I have suggested ways to avoid cyclic and unproductive imagining. I have also shown that here-and-now sensory experience is essential to vital imagination. But it is also extremely important to use imagining *to do something.* Idea-sketching is an important kind of doing that clarifies and records inner imagery, links imagining with seeing (by making ideas visible), and adds an element of action to thinking itself. As you will see in the next section, idea-sketching fills in especially where seeing and imagining are limited, making visual thinking more effective.

References

1. Maltz, M. *Psycho-cybernetics.* Wilshire Books.
2. De Mille, R. *Put Your Mother on the Ceiling.* Walker.
3. Bruner, J. *On Knowing: Essays for the Left Hand.* The Belknap Press of Harvard University Press.
4. Kelly, T. *Testament of Devotion.* Harper & Row.
5. Tyrell, G. *The Personality of Man.*
6. James, W. *Psychology: The Briefer Course.* Harper & Row (Harper Torchbooks).
7. Bartlett, F. *Remembering.* Cambridge University Press.
8. Miller, G., Galanter, E., and Pribram, K. *Plans and the Structure of Behavior.* Holt, Rinehart & Winston.

IDEA-SKETCHING

Idea-sketching is the faculty of visual thinking that gives birth to ideas. The following three chapters treat the expressive nature of idea-sketching, the importance of fluency in graphic language to the development of visual ideas, and the use of strategies to stimulate fuller idea exploration.

ETC: EXPRESS/ TEST/CYCLE

20

Graphic Ideation

An idea is an object of the mind. I look at my hand. My hand is not an idea, but my perception of my hand is. I look at the words that I am currently putting down on paper; these squiggly marks are not ideas, but the images and verbal concepts now running through my head are. Ideas are internal constructs of perception, imagination, and thinking.

To "ex-press" means to "press out." Idea-sketching is a way to express visual ideas, to literally press them out into tangible form. Visual ideas can be expressed by acting them out, talking about them, writing them down, constructing them directly into a three-dimensional structure—and drawing them. In this section, I will concentrate on one form of idea-sketching: *graphic ideation*, or generating and expressing ideas by means of drawing.

Graphic ideation has two basic modes: exploratory and developmental. In the exploratory mode, the visual thinker uses drawing as a means of probing his or her imagination, seeking to touch and record the vague and elusive imagery that usually accompanies the conception of a new idea (see Chapter 18). In the developmental mode, the visual thinker gradually evolves a promising, though initially embryonic, concept into mature form.

In the exploratory mode of graphic ideation, thinking and sketching are adventurous. Consider yourself to be flying over the terrain of your problem in a highly maneuverable "mental helicopter": climbing for an overview, rapidly descending to record an interesting possibility, and then climbing to seek another viewpoint. Each sketch captures general features only, not details; it is a kind of rough map that will enable you to return later to the concept, if you choose to develop it further. From the vantage point of your maneuverable mental helicopter, you can readily see many alternative concepts: exploratory sketching is therefore rapid-fire and copious.

In the developmental mode of graphic ideation, you land your mental helicopter and proceed on foot to develop a more thorough understanding of a promising concept. Now the emphasis is on one idea, and the pace is more deliberate and thorough. Developmental sketching is less schematic and more concerned with concrete details. No less creative than the exploratory mode, developmental graphic ideation is, in fact, a focused form of exploration.

The exploratory and developmental processes of graphic ideation are different from those involved in "sketching from life." In the section on seeing, you drew from a model that was visible, fully formed, and accessible for prolonged and repeated viewing. By contrast, *the model for idea-sketching is an inner event visible only to the mind's eye, rarely fully formed, and easily lost to awareness.* Visual thinkers who use drawing to explore and develop ideas make *many drawings:* idea-finding and formation is not a static, "one-picture" procedure. They also *draw quickly:* ideas rarely hold still; they readily change form and even disappear. In both the exploratory and the developmental mode, graphic ideators also use *many graphic idioms.* When you are sketching from life or communicating a visual idea to others, you can be content with one graphic idiom. But when you are exploring ideas, you must use graphic language more flexibly, as the next chapter shows.

The quickly executed, formative processes of graphic ideation are favored by sketching. What are the characteristics of a sketch? Actors perform theatrical "sketches" that are customarily short and informal; writers "sketch out" their ideas in outline form and in rough, preliminary drafts; sculptors make rapidly executed "three-dimensional sketches" before proceeding to the final expression of their idea. In whatever form it takes, a sketch is typically (1) self-intended or directed to a small in-group, (2) concerned more with chief features than with details, and (3) performed spontaneously and quickly. Also, sketches that record the excitement of generating and formulating ideas often possess a vitality and freshness lacking in the final communication.

As suggested in Chapter 2, graphic ideation is not to be confused with graphic communication. The former is a formative process concerned with conceiving and nurturing ideas; the latter is an explanatory process concerned with presenting fully formed ideas to others. *Graphic ideation is visually talking to oneself; graphic communication is visually talking to others.* Graphic ideation precedes graphic communication in most instances: the purpose of graphic ideation is to discover and develop an idea worth communicating.

In graphic ideation, being your own audience, you enjoy certain freedoms denied the graphic communicator: you can sketch freehand, quickly and spontaneously, leaving out details that you already understand or that you believe might concretize your thinking prematurely; you can use whatever graphic idiom furthers your thinking, without concern that others be able to understand it; you can feel free to fail many times on the way to obtaining a solution. The upper drawing in Figure 20-1, from the notebook of Thomas Alva Edison, is an example of graphic ideation, a visible step in the birth of an idea. The lower drawing, from a General Electric catalog, is an example of graphic communication, a presentation to others of an idea already fully formed.

A Feedback Loop

Graphic ideation utilizes seeing, imagining, and drawing in a cyclic feedback process. I have given this feedback loop the acronym ETC (etcetera) to dramatize the importance of repetitive cycling to the graphic development of visual ideas.

The diagram in Figure 20-2 illustrates how ETC works. Note first the "arrow in" and the "arrow out": these represent an input and an output of information. An input of information is typically a statement of a problem and information relative to the problem; an output is usually a communication of a solution. The first step in graphic ideation is the expression of ideas (Express). Next, ideas are carefully evaluated (Test). Then the thinker returns to another round of idea expression with information gained in testing and frequently with another strategy for generating ideas (Cycle). The ETC loop in the center of the diagram presumes information in and excludes information out; *graphic ideation, by means of ETC, is concerned primarily with processing, not with collecting or communicating, information.* In other words, graphic ideation is the idea factory; it is not the supplier of raw materials and not the marketer of the finished product.

The basic process of graphic ideation is to Express, then Test, then Cycle, ETCetera, ETCetera, until the desired idea is fully conceived, at which point the final letter is E: Express. For reasons that I will now develop, expression, testing, and cycling should be treated as distinct and sequential steps in a developmental process, and not as steps that can be performed simultaneously.

Expression of Ideas

Most beginners in graphic ideation find the expression of ideas to be the most difficult part of ETC. The paper is menacingly blank; imagination falters; whatever is expressed in a sketch comes out differently than intended. In the next few paragraphs, I will treat four basic prin-

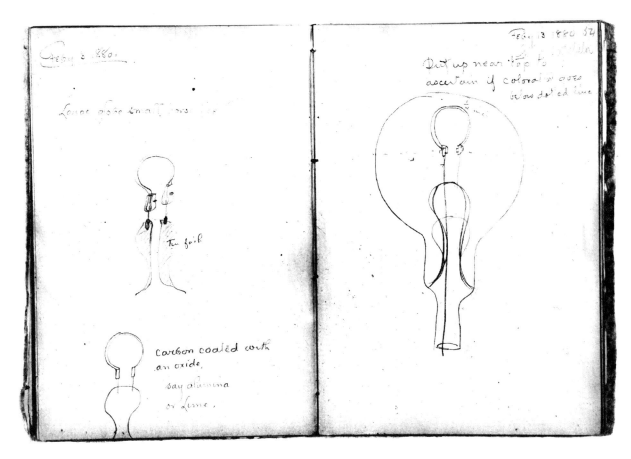

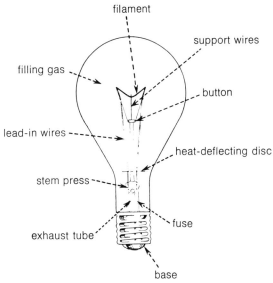

filament

support wires

filling gas

button

lead-in wires

heat-deflecting disc

stem press

fuse

exhaust tube

base

Figure 20-1.

J. P. Guilford, a pioneer in the psychology of creativity, suggests the idea-releasing principle of *fluent and flexible ideation.*[1] Fluent ideation is demonstrated by a thinker who generates many ideas; the yardstick of fluency is quantity—not quality or originality. Flexible ideation is demonstrated by a thinker who expresses diverse ideas; the measure of flexibility is variety.

The next exercise (20-1), taken from a psychological test, will enable you to evaluate how fluently and flexibly you currently generate and express ideas.

ciples intended to help open the flow of ideas onto paper: (1) fluency and flexibility of ideation; (2) deferred judgment; (3) unhesitating response; and (4) skill in drawing.

20•1 THIRTY CIRCLES

1. On newsprint, draw 30 freehand circles, each approximately 1″ in diameter and 2″ apart.

2. In 5 minutes, draw a few sketchy details into or around each circle to make it an identifi-

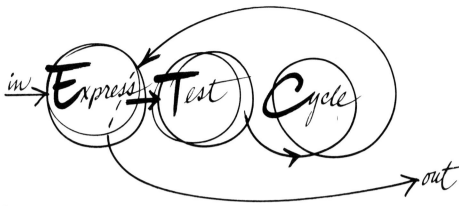

Figure 20-2.

able image (such as the planet Venus, a baseball, or a teapot). Work rapidly; fill one circle every 10 seconds and you will complete all 30 in 5 minutes.

3. How fluent in graphic ideation were you? Did you complete all 30 circles? How flexible were you? Did you fall into ruts (such as drawing four faces), or did you represent a variety of ideas?

A second idea-generating principle is *deferred judgment*. Attempting to express and to judge ideas simultaneously is like trying to drive a car with one foot on the accelerator and the other on the brake. Alex Osborn, the inventor of brainstorming, found this interesting instruction in a letter written by Friedrich Schiller to "a friend who complained that he was unable to generate ideas":

The reason for your complaint lies, it seems to me, in the constraint which your intellect imposes upon your imagination. . . . Apparently it is not good—and indeed it hinders the creative work of the mind—if the intellect examines too closely the ideas already pouring in, as it were, at the gates. . . . In the case of the creative mind, it seems to me, the intellect has withdrawn its watchers from the gates, and the ideas rush in pell-mell, and only then does it review and inspect the multitude. You worthy critics, or whatever you may call yourselves, are ashamed or afraid of the momentary and passing madness

which is found in all real creators. . . . Hence your complaints of unfruitfulness, for you reject too soon and discriminate too severely.[2]

Schiller gave this valuable advice in 1788; the importance of deferred judgment to unfettered idea generation has been long recognized. Bear it in mind as you practice visual brainstorming in exercise 20-2.

20·2 VISUAL BRAINSTORMING

Brainstorming is an idea-generating activity that can be performed individually or in a group. Although brainstorming is usually verbal, it can also take other forms: a dancer can brainstorm directly in dancing, a composer can brainstorm musically—and a visual thinker can brainstorm visually. Visual brainstorming is a basic strategy for exploratory graphic ideation, and also a sure-fire remedial strategy whenever thinking has become stale or stuck. Its two basic principles are:

1. *Defer judgment.* (Whether brainstorming solo or with others, don't criticize ideas until the brainstorming session is over.)

2. *Reach for quantity.* Take a simple problem that interests you. Generate a series of thumbnail idea-sketches on that problem. Set a quantity goal (say 30 idea sketches in 60 minutes). Also keep a tally of every time you find yourself judging an idea while brainstorming.

William James wrote that "whenever a movement unhesitatingly and immediately follows upon the idea of it, we have ideo-motor action. We think the act and it is done."[3] Most habitual behavior follows this description: I have the idea that I would like another sip of coffee—no sooner thought than done, as my hand quite automatically reaches for the cup. More complex performances, such as piano playing, illustrate the same phenomenon: accomplished pianists experience no lapse between their awareness of a musical idea and a corresponding motion of their fingers across the keyboard. The flow from idea to expressive motor action is *unhesitating* and *immediate*.

The quick and spontaneous release of idea into sketch is especially important to graphic idea expression. For one thing, new and undeveloped ideas are ephemeral: Graham Wallas tells of a man who had so marvelous an idea that he rushed into his garden to thank God for it; upon rising from his knees, he realized he had forgotten the idea. Better to draw first and pray later! Ideas also frequently appear in rapid succession. *Unhesitating sketching response to each idea that arises creates a momentum in which expression keeps pace with thinking.* Finally, immediate graphic response to each idea helps prevent the intervention of conscious, judgmental processes. The hyphen in the term "ideo-motor action" represents a natural and spontaneous flow that is easily blocked by the intrusion of judgment—or of another idea.

Unhesitating expression of visual ideas into sketch form can be encouraged by keeping an "idea log" (20-3).

20•3 IDEA LOG

An idea log can take many forms. You may find that sketches on separate sheets of paper or file cards are easier to compare and to group in the Test phase of ETC. Or you may find that sketching on long scrolls of paper encourages idea flow. Accordion-folded scrolls combine book and scroll features and, in the form of used computer paper, can often be obtained free. Whatever form it takes, keep your idea log always with you (even by your bed, ready for recording an insight gained in a dream).

Fluent and flexible ideation, deferred judgment, and unhesitating translation of idea into sketch are important ways to open the gates that hold back ideas. However, the importance of *drawing skill* to the full expression of visual ideas must not be overlooked. Inadequate drawing ability has three negative effects on the Express phase of ETC: (1) a clumsy sketch can evoke judgmental processes that restrict or stop idea flow; (2) ideas that cannot be adequately recorded in sketch form are often lost; and (3) attention devoted to problems of drawing is attention diverted from generating ideas.

Unfortunately, there is no shortcut to drawing skill. Return to Chapter 5: have you found drawing materials that give you pleasure to use and that do not incur frustration? Have you developed a well-organized working environment? Return to Chapter 6: idea-sketching is best performed in a state of relaxed attention. Return to Chapter 10: practice in making thumbnail sketches is especially relevant to idea-sketching. Return to Chapters 15 and 16: drawing from memory and making sketches of your dreams will give you practice in recording inner imagery. You would not expect to be able to express your ideas verbally without verbal language skill acquired by many years of schooling and practice. Similarly, do not expect satisfying visual idea expression without well-informed and sustained practice in the use of graphic language. As William Lockard says: "If you would learn to draw, hold the [drawing] instrument often—and hold it with your head."[4]

Time for Testing

Once you have expressed a number of ideas in sketch form, you are ready to evaluate them. Judgment, deferred in the Express phase of ETC, is fully exercised in the Test phase. *Now* is the time to be self-critical, not before. Testing involves (1) *seeing* your sketches fully and imaginatively, (2) *comparing* sketches, (3) *evaluating* each idea in relation to present criteria, and (4) *developing new criteria.*

Idea sketches are a remarkable extension of imagination, a kind of visible graphic memory. Like actual memory images, however, idea sketches are useful only to the extent that they are accessible. An idea sketch that is tucked away in a file or lost is comparable to a memory image that cannot be recalled. A missing sketch is the victim of "graphic amnesia." The first step in the Test phase of ETC is to display all of your idea sketches side by side (20-4). Once displayed, your graphic memory is fully available for the active operations of testing (20-5).

In Figure 20-3, William Katavolos and his students are fully enveloped in, and have full visual access to, a collective graphic memory of ideas generated on a design problem.

20•4 DISPLAY YOUR GRAPHIC MEMORY

Place all of your idea sketches on a wall, table, or floor. Step back for an overview.

As you view your idea sketches, attempt to see them as fully and imaginatively as possible, recentering the way you see into a variety of viewpoints.

20•5 RECENTER

1. The most crucial imaginative act in the Test phase is moving from the viewpoint of creator to the viewpoint of critic. As you view your sketches, imagine yourself in the role of a constructively critical person who is seeing them for the first time.

2. In the role of constructive critic, try other modes of recentering described in Section III. Turn some of the sketches upside down; critically look at the gestalt of the entire display; project alternative images on specific sketches.

Figure 20-3.

Still in the recentered viewpoint of critic, compare your sketches (20-6). As you do, notice another important characteristic of fluent and flexible ideation. In addition to freeing the flow of idea expression, numerous and varied sketches provide valuable material for comparison. Also note how the format of your idea log influences your ability to compare. A bound notebook makes comparison clumsy; a continuous scroll of sketches prevents side-by-side comparison. Comparison, essential to the act of evaluation, is facilitated by a loose-leaf format that permits you to juxtapose and group your sketches freely.

20•6 COMPARE

In the spirit of wine-tasting (see Chapter 11), compare austere ideas with the most complex, and compare each idea in relation to your criteria. Physically grouping and regrouping the sketches usually facilitates comparison. Moving your sketches out of the order in which they were expressed and into new juxtapositions also may help you see ideas afresh.

As you compare your idea sketches, you automatically begin to evaluate them. Make written notes and sketches *immediately* to catch the essence of each evaluation. Use a different-colored marker (some like red) to denote the Test phase (20-7). The act of putting your evaluations into writing enforces a thorough review, reinforces a critical viewpoint, and assures that judgments are not forgotten.

20•7 COLORED NOTATIONS

Use a different-colored marker to record evaluations, comments, and ideas obtained in Testing.

Testing, of course, implies criteria. In the early rounds of ETC, criteria are usually imprecise, incomplete, and implicit. Initial criteria are also frequently inaccurate. The final function of the Test phase is to review criteria and to state them more exactly. *Try using your idea sketches to question earlier judgments about criteria.* For example, have an idea sketch "ask" whether this or that criterion is as important as you first deemed it to be or whether you have overlooked an important criterion (for example, is something missing?). As you formulate and refine your criteria, record them in writing (20-8). A revised statement of criteria is an invaluable aid in the next round of ETC.

20•8 CRITERIA FORMULATION

At the end of the Test phase, formulate your new understanding of problem criteria in writing.

Can you eliminate a criterion? As you look at your idea sketches, can you detect criteria that you are currently using but have not yet listed? Can you reprioritize your criteria? Can you state them more accurately or more succinctly? Rejuggling and restating criteria is an excellent way to clarify your thinking as you enter the Cycle phase of ETC.

Now Cycle

The first round of idea-sketching rarely produces an idea that fully meets your test. After evaluating your first concepts, you are ready to return to idea-sketching. At this point, it is often valuable to pause and consider the next strategy you will use in search of a solution. *Cycling, the third step in ETC, is more than a return to another round of idea expression: it is a return with an idea-generating strategy in mind.*

What is a strategy for graphic ideation? One strategy might be to develop a particular concept in considerable detail; another might be to generate more ideas before delving into detail with one. For most people, such choices are rarely choices at all; instead, they are robot-like steps, the unconscious product of habitual patterns of problem-solving. The primary value of cycling in ETC is the emphasis it puts on choice. Cycling asks that you *consciously* choose your next idea-generating strategy (20-9). To abandon habitual patterns, to think in an active, voluntary way: what could be more central to independent, creative thinking?

20•9 CYCLING

A number of strategies for graphic ideation are described in Chapter 22. Select the strategy that you believe will most likely help you into the next round of ETC.

Once you have chosen a strategy and enter again into the Express phase of ETC, remember the principle of deferred judgment. Do not attempt to judge whether the strategy is working: focus your attention upon idea generation. You will have ample opportunity to assess the value of the chosen strategy in the next Test phase.

Develop Your Own Style

At first, ETC may seem mechanical and unnatural. Like any skilled behavior, ETC is clumsy and difficult until it has been practiced for a period of time. If you persevere, however, you will develop facility in idea expression, rigor in evaluating your own ideas, and an enlarged repertoire of idea-generating strategies. Finally, you will develop a style of graphic ideation that is distinctly your own.

The exploratory and developmental processes of graphic ideation can be expressed in many ways. Paul Klee wrote: "A certain fire flares up; it is conducted through the hand, flows to the picture and there bursts into a spark, closing the circle whence it came, back into the eye and farther."[5] And Edward Hill observed that "generally the process is one of evolution. Blurred mental images, once projected onto paper, are immediately given a new identity by line. The original idea may now be judged by the eye, developed, and resolved if resolution proves possible— altered or discarded if it does not."[6] E, T, and C are intended merely to remind you of the basic structure of an idea-formulating process that you will, with practice, form and refine into a form-giving act that is uniquely your own.

"Any ideas I have immediately become concrete in sketches," says film writer and director Federico Fellini. "Sometimes the very ideas are born when I'm drawing."[7]

References

1. Guilford, J. *The Nature of Human Intelligence.* McGraw-Hill.
2. Schiller, F. Quoted by Osborn, A., in *Applied Imagination.* Scribner's.
3. James, W. *Psychology: The Briefer Course.* Harper & Row (Harper Torchbooks).
4. Lockhard, W. *Drawing as a Means to Architecture.* Reinhold.
5. Klee, P. *The Thinking Eye.* Wittenborn.
6. Hill, E. *The Language of Drawing.* Prentice-Hall (Spectrum Books).
7. Fellini, F. Quoted by Hamblin, D. J., in "Which Face Is Fellini?," *Life Magazine,* July 30, 1971.

OUT OF THE LANGUAGE RUT

21

Many Graphic Languages

Human beings communicated with pictures long before they developed the kind of nonpictographic language that you are currently reading. Pictographic symbols record various degrees of abstraction. The cave painting in Figure 21-1 symbolizes a specific animal; the Egyptian hieroglyph and the more abstract Chinese calligraph symbolize "man"; the American Indian circle-symbol represents the highly abstract idea of "all."

Although contemporary alphanumeric languages evolved from pictographic languages, we should not infer that words and numbers are necessarily more recent or advanced. Perspective and orthographic projection, for example, postdate the alphabet by many centuries; a recent advance in computer technology is the graphic computer. The computer presentation in Figure 21-2 is an example of a contemporary application of graphic language: it is an isometric diagram of population distribution in the United States, developed by geographers Waldo Tobler and Frank Rens. Many professions cur-

rently use graphic languages: physicists draw diagrams and graphs, business managers draw organization charts, and physiologists draw cross-sections. Indeed, you will find graphic-language expressions on the blackboards of almost every department of a university, from aeronautics to zoology. Visual and verbal-mathematical modes of thinking and communicating are complementary: one is not higher than the other.

The basic purpose of this chapter is to encourage literacy in many graphic languages as a means toward thinking flexibly. *Thinkers who have a broad command of graphic language not only can find more complete expression for their thinking but also can recenter their thinking by moving from one graphic language to another.* Unlike verbal thinkers who usually learn only one language, visual thinkers can very easily acquire a variety of graphic languages, some ancient, others modern, some abstract, others concrete. By changing languages, they can largely avoid the "language rut" that holds thinking to a fixed viewpoint and a limited set of mental operations.

Thought and Language Interact

What is the relationship between visual thinking and graphic language? Vygotsky writes that "schematically we may imagine thought and speech as two intersecting circles. In their overlapping parts, thought and speech coincide to produce what is called verbal thought."[1] By the same analogy, visual thinking and graphic language also interact in graphic ideation, as illustrated in Figure 21-3.

The overlapping circles in Figure 21-3 dramatize two important observations about the interaction of thinking and language. First, not all visual thinking is language thinking: visual thinking can utilize operations (such as the act of synthesis), can be represented by imagery (such as perceptual and mental imagery), and can occur at levels of consciousness (such as dreaming) outside the realm of language thinking. Second, not all use of graphic

Figure 21-1.

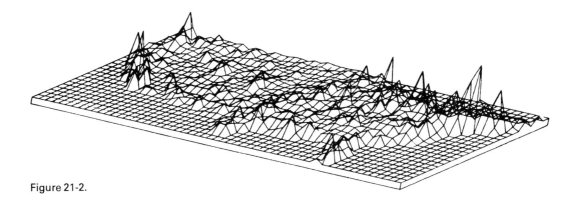

Figure 21-2.

language involves thinking: a major use of graphic language is to communicate the *result* of thinking to other people.

The actual neurological interaction between thinking and language is extremely complex and little understood. The overlapping circles in Figure 21-3 *stand for* this reality, or more correctly, stand for a mental concept of this reality. In this regard, the two circles are similar to any graphic statement. A perspective sketch of a building is not the building itself; the actual building is far more than the sketch. Graphic symbols, whether abstract or concrete, are always less than they represent.

Further, a given graphic symbol represents only one of many ways to view an idea. C. K. Ogden and I. A. Richards, for example, chose to illustrate their notion of the relation of thought and language not with circles but with the triangle shown in Figure 21-4.[2]

Notice that the triangle allowed Ogden and Richards to introduce a third element, the "referent" (the actual thing or event to which thinking or symbolization refers) and also the notion that a language symbol stands for, but is not the same as, the referent. Notice also that Vygotsky's overlapping circles and Ogden and Richards' triangle stand for essentially the same concept, but cause us to view that concept somewhat differently.

Similarly, the perspective sketch of a car in Figure 21-5 represents only one way of looking at a car. We could as well view the car orthographically, from above or below, in cross-section, in a schematic drawing that shows its electrical system, in a

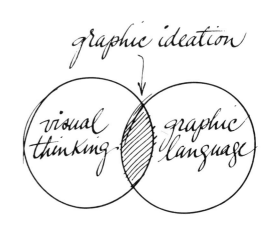

Figure 21-3.

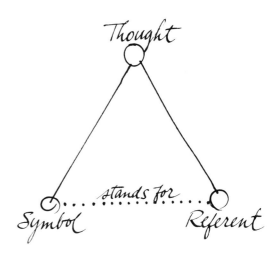

Figure 21-4.

Figure 21-5.

diagram that shows it as an abstract element in a larger transportation or ecological system, and so on. Each graphic expression would be less than the actual car, and each would represent only one way of looking at the car. Taken together, however, they would form a composite picture more real than the "realistic" drawing in Figure 21-5.

Translating thinking into graphic language also involves the visualizer in a certain set of mental operations. The person who drew the orthographic views in Figure 21-6, for example, was literally forced to consider the concept in detail and true proportion; he or she also necessarily rotated the form and cut through its structure. On the other hand, the person who drew the diagram in Figure 21-7 was directed, by the operations inherent in this more abstract graphic language, to consider overall relationships and to ignore specific and concrete details of structure. *By moving from one graphic language to another, visual thinkers automatically apply the built-in mental operations of each language. In effect, they use language to expand the range of their thinking.*

What have I said so far about the relation of visual thought and graphic language? First, not all visual thinking is language thinking. Second, a graphic symbol is always less than what it represents; "the word is not the thing" and the graphic symbol is not either. Third, every graphic expression embodies a viewpoint, a single

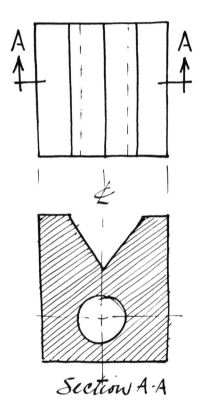

Figure 21-6.

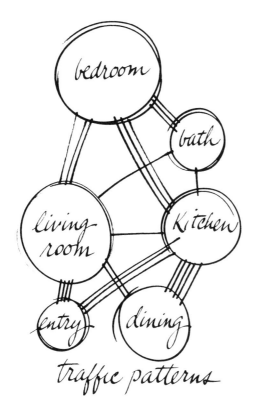

traffic patterns

Figure 21-7.

way of looking at reality; by encoding an idea in a variety of graphic languages, the visual thinker represents the idea more completely. Fourth, every time the thinker changes graphic languages, he or she submits the idea to a new set of built-in mental operations.

To be sure, the relation of thinking to language poses a number of pitfalls. Embryonic ideas can readily be deformed in the act of expression into language: lack of language skill or inappropriate choice of language can be especially damaging to tender new concepts. The thinker is also easily tempted to mistake graphic images for the reality that they represent: a science student who graphically represents a magnetic field as a pattern of lines may be misled to believe that magnetism is actually linear; the designer who has glamorized an idea in a drawing may come to believe in this graphic illusion. Further, language can enable the thinker to conceal what needs to be revealed: an architect may be reluctant to choose a perspective viewpoint that exposes a "bad side" or unsolved problem; he or she may also persist

in the use of perspective to consider external appearance, when in fact another graphic language should be used as a tool to investigate internal structure or human activities inside the building. Finally, the thinker who habitually uses only one or two graphic languages is often led by language to avoid the kinds of mental operations that are most needed to solve the problem.

In short, the relation between visual thinking and graphic language can be both beneficial and harmful. To use graphic language as a tool for thinking, and at the same time to avoid the pitfalls and ruts of language, the visual thinker must learn to use graphic language *flexibly.*

Graphic Language Flexibility

Language, writes Jerome Bruner, "predisposes a mind to certain modes of thought. . . . Also, languages differ in their capacity to absorb and facilitate new ideas."[3] Ernst Cassirer elaborates: "Each language draws a magic circle around the people to which it belongs, a circle from which there is no escape save by stepping out of it into another."[4] *Visual thinkers are fortunate that they have relatively easy access to a variety of graphic languages; if one language does not facilitate their thinking, they can step out of it and into another.*

The ability to use language to express a wide range of subjective-to-objective meanings is one kind of graphic-language flexibility. Picasso claimed, "What I want is that my picture should evoke nothing *but* emotion."[5] Compare Picasso's idea sketch for his famous painting *Guernica* in Figure 21-8 with some of the more objective sketches reproduced in this chapter, and with your own sketches. In exercise 21-1, how able are *you* to move freely between subjective and objective graphic expression?

21·1 EXPRESSIVE FLEXIBILITY

Edward Hill observed that "Such is the physiognomy of a student's drawing that it reveals the temperament . . . of its author. By temperament we mean his timidity or temerity, dullness or sagacity, frivolity or earnestness."[6]

1. With a teacher or friend, look over a collection of your recent drawings. Can you diagnose a quality of emotional expression in them that begs to be loosened up, to be made more flexible? For example, are most of your sketches tight, dry, and over-concerned with objective description? Or are they subjectively oriented and short on logical, careful concern for structure?

2. Devise an experience intended to help you achieve more flexibility of expression.

Freedom to break out of established graphic languages and to create new ones is another important kind of graphic-language flexibility. Usually the need for this kind of flexibility is experienced in relation to a language that is felt to be constricting. Perspective, for example, was invented to free Renaissance artists from the limitations of medieval language forms, and Cubism was invented to release contemporary artists from the limitations of perspective. Today, many designers dislike the limitations on spatial thinking imposed by languages based on Cartesian coordinates. When a non-Cartesian language is developed that successfully unhinges thinking from T-square and triangle, design thinking (and the quality of our designed environment) will be substantially changed.

Of course, someone who is not yet fully acquainted with the advantages and disadvantages of current graphic-language forms is hardly prepared to invent new ones. Inventiveness in graphic language requires a modicum of graphic literacy. A way to gain literacy, and at the same time to acquire language flexibility, is to learn how to use graphic language to move thinking and expression from abstract to concrete meanings and back. Indeed, the ability to move from one graphic language to another, along the dimension of abstract-to-concrete, is probably the most useful kind of graphic-language flexibility.

Figure 21-8.
Pablo Picasso. *Horse and Bull*, early May, 1937. Study for *Guernica*. Pencil on tan paper, 8⅞" × 4¾". On extended loan to The Museum of Modern Art, New York, from the artist.

Abstract-to-Concrete

In this book, you have experienced the dimension of abstract-to-concrete in a number of contexts. In the context of seeing, you have witnessed your tendency to seek overall visual patterns and then to analyze these gestalten in detail. In the context of imagining, you have glimpsed abstract inner imagery and the more vivid and representational imagery that accompanies dreams and visual recall. In the context of drawing, you have drawn abstract thumbnail sketches and then developed specific spatial features in perspective drawings. The dimension of abstract-to-concrete pervades all cognitive activity; as Ulric Neisser puts it, all cognition consists of a two-stage act of construction: "the first is fast, crude, wholistic, and parallel, while the second is deliberate, attentive, detailed, and sequential."[7]

Mental activities of abstraction and concretization, by the interaction of thought and language, have naturally been embodied into language form. S. I. Hayakawa demonstrates the abstract-to-concrete dimension of verbal language with the "abstraction ladder" shown in Figure 21-9. At the bottom of the ladder is reality—not the reality that you can perceive (that is Step 2) but the reality of whirling atomic particles and energy processes that cannot be perceived even with the finest of scientific instruments. As you

proceed up the abstraction ladder, each word is more abstract—that is, it makes less reference to a particular thing or event. As Hayakawa points out, "This process of abstracting, or leaving things out, is an indispensable convenience."[8] An abstract word such as "wealth" allows thinking to embrace far larger meanings than does the concrete word "Bessie."

In Figure 21-10, I have organized graphic languages in a hierarchy of abstraction, much like Hayakawa's verbal one. In the next few pages, I will describe and provide examples of each of these kinds of graphic language. Notice, in Fig-

ABSTRACTION LADDER
Start reading from the bottom *UP*

8. "wealth"

8. The word "wealth" is at an extremely high level of abstraction, omitting *almost* all reference to the characteristics of Bessie.

7. "asset"

7. When Bessie is referred to as an "asset," still more of her characteristics are left out.

6. "farm assets"

6. When Bessie is included among "farm assets," reference is made only to what she has in common with all other salable items on the farm.

5. "livestock"

5. When Bessie is referred to as "livestock," only those characteristics she has in common with pigs, chickens, goats, etc., are referred to.

4. "cow"

4. The word "cow" stands for the characteristics we have abstracted as common to cow_1, cow_2, cow_3 . . . cow_n. Characteristics peculiar to specific cows are left out.

3. "Bessie"

3. The word "Bessie" (cow_1) is the *name* we give to the object of perception of level 2. The name *is not* the object; it merely *stands for* the object and omits reference to many of the characteristics of the object.

2.

2. The cow we perceive is not the word but the object of experience, that which our nervous system abstracts (selects) from the totality that constitutes the process-cow. Many of the characteristics of the process-cow are left out.

1. The cow known to science ultimately consists of atoms, electrons, etc., according to present-day scientific inference. Characteristics (represented by circles) are infinite at this level and ever-changing. This is the *process level*.

ure 21-10, the line that demarks abstract graphic languages from concrete ones. The dichotomy that mistakenly links verbal thinking with abstraction and visual thinking with concretization was undoubtedly conceived by someone who identified visual imagery with postcard realism and failed to observe visual abstraction as expressed in contemporary art and abstract graphic-language forms. Abstract graphic languages encode abstract ideas, not concrete things. By distinguishing abstract from concrete graphic languages, I hope particularly to dramatize the abstract dimension of graphic-language thinking.

Abstract Graphic Languages

In an important sense, all graphic symbols are abstract—that is, all leave out part of what they represent; one cannot draw without being selective. However, some graphic symbols are far more abstract than others. The "yin and yang" symbol in Figure 21-11, for example, refers to the abstract notion that the interplay of contrasting and complementary forces engenders and sustains the universe. Such a symbol represents an abstract concept, not a thing, and is comparable to an abstract word.

Figure 21-11.

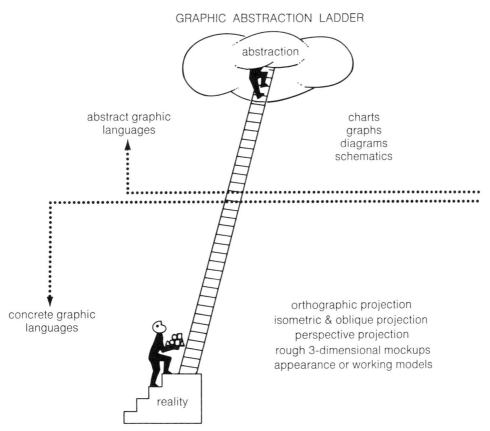

Figure 21-10.

A single graphic symbol, however, does not constitute a graphic language. *A language consists of a set of rules by which symbols can be related to represent larger meanings.* A simple example of an abstract graphic language is the *Venn diagram* illustrated in Figure 21-12 (and also in Figure 21-3). The basic grammatical rule of the Venn diagram is that overlap means relationship. When the Venn diagram is used to illustrate principles of logic, additional rules are added—shading, for example, describes *kinds* of relationship. As with all language forms, the Venn diagram can articulate a large variety of meanings. Additional circles can be added, and the rules are not changed when the overlapping symbols are not circles.

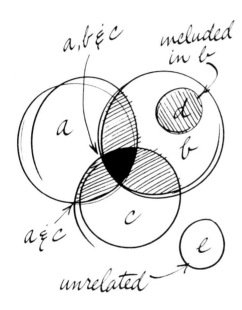

Figure 21-12.

Another abstract graphic-language form is the *organization chart* in Figure 21-13. An organization chart can represent many things: a family tree, a management hierarchy, a system for making decisions, and so on. (A language can always encode a variety of contents.) The grammatical rules of this particular language are (1) the higher an item is on the page, the more important it is, (2) equal rank is positioned on the same horizontal level, (3) like functions are grouped together, and (4) lines

represent connectedness (of genes, of power, of information). An organization chart is an abstract language because it helps to describe the structure of an idea, not a thing.

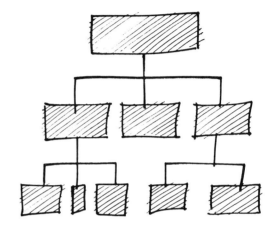

Figure 21-13.

The *flow chart* in Figure 21-14 is an elaboration of the organization chart. In this language, the arrow symbolizes direction of flow between elements. Notice that the elements in this case are no longer boxed words but highly abstract representations.

Pattern language was invented by architects Christopher Alexander, Sara Ishikawa, and Murray Silverstein to aid in the design of buildings. Pattern language consists of a number of abstract visual symbols comparable to newly conceived words. A typical symbol, shown in Figure 21-15, visually represents a desired relationship or attribute; in this example, the symbol represents the designers' desire that "windows near places where people spend more than a minute or two should all look out on areas of 'life.' "[9] This pattern (which is supported by several pages of documented research) represents one of many patterns that are also given an identifying symbol. Figure 21-16 shows how several patterns are clustered together in a "language cascade." The syntax of pattern language requires that (1) patterns that influence other patterns are placed higher on the page, (2) related patterns of equal

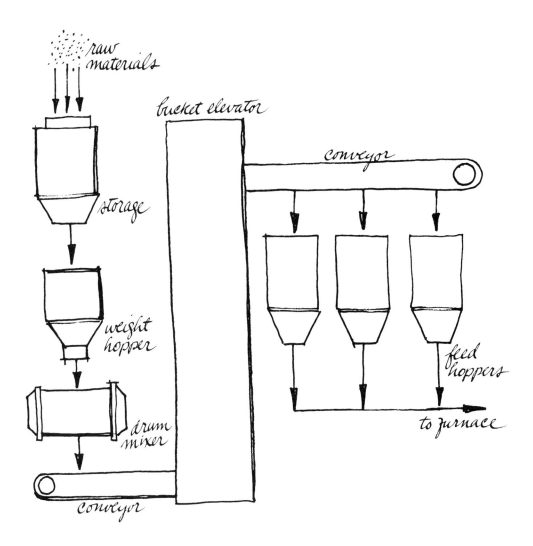

Figure 21-14.

importance are grouped together horizontally, and (3) connecting arrows denote only important relationships. The abstract, visual, and flexible nature of pattern language makes it extremely useful as a device to express, to see, and to think about complex problems (21-2).

Figure 21-15.

Figure 21-16.

21·2 PATTERN LANGUAGE

1. On the blank side of separate file cards, describe each basic requirement of a given problem with an abstract graphic symbol. On the back, elaborate the requirement in writing.

2. Now place all of the file cards on a large table with graphic symbols up. Manipulate the symbolized requirements boldly, grouping and regrouping them into various relationships.

3. Using the grammar of pattern language, record the most idea-provoking arrangements.

Still another abstract graphic language is the *link-node diagram*, shown in the form of a "sociogram" in Figure 21-17. Here, direction and importance of relationships, or links, between people are represented by coded arrows.

The *bar chart* (top) and the *graph* in Figure 21-18 are perhaps more concrete than the previously described abstract graphic languages because they encode specific dimensions. Also more concrete are the *schematic* drawing and circuit *diagram* in Figure 21-19, which describe functional and spatial relationships in nondimensional terms.

The following few pages contain examples of idea-sketching in various abstract graphic languages. Notice how *the use of abstract graphic languages fosters fluent ideation, encourages bold manipulation of basic relationships, and generates many concrete alternatives.*

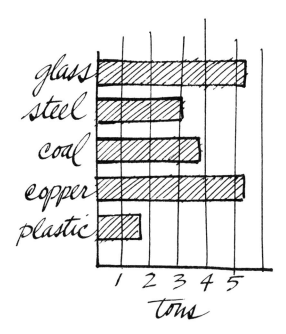

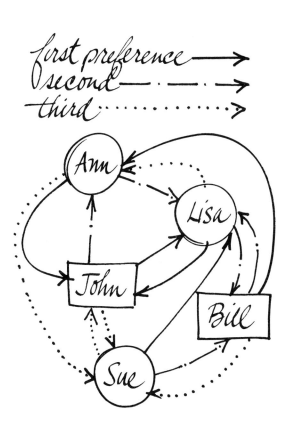

Figure 21-17.

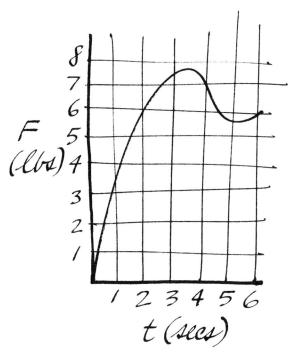

Figure 21-18.

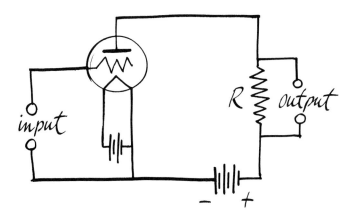

Figure 21-19.

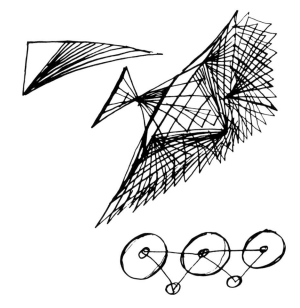

Figure 21-20. President Herbert Hoover's doodle does not represent a thing, but expresses a reverie, a feeling, or an aesthetic interest in structure. It is therefore abstract in the same sense that nonrepresentational art is abstract.

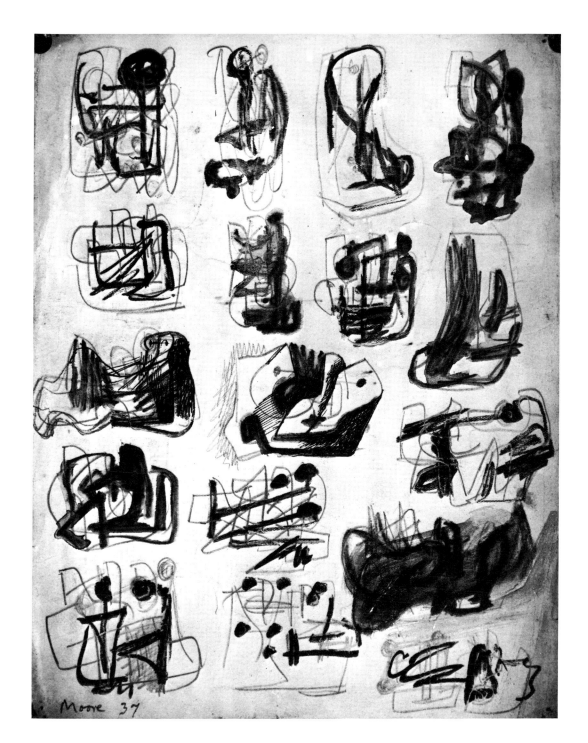

Figure 21-21. Henry Moore's idea-sketches are concerned with the gestalt of a sculptural idea, with bold patterns of light and shadow, with overall relationships of line, volume, and space; abstraction permits him to explore many ideas quickly. Moore says that he sometimes begins sketching with nothing in mind, and then discovers an idea in his doodlings. (See exercise 10-5, Da Vinci's Device.)

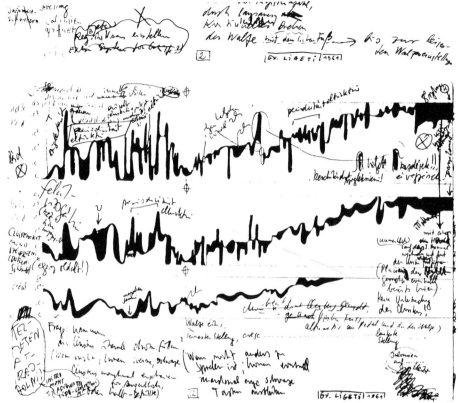

Figure 21-22. György Ligeti, like many contemporary composers, finds traditional musical notation unsuited to the expression of his ideas. In this notation for his score for *Volumina,* Ligeti invented an abstract graphic language to represent ideas that are essentially nonvisual.

Copyright © by Henry Litolff's Verlag, care of C. F. Peters Corporations, 373 Park Avenue South, New York, N.Y. 10016. Reprint permission granted by the publisher.

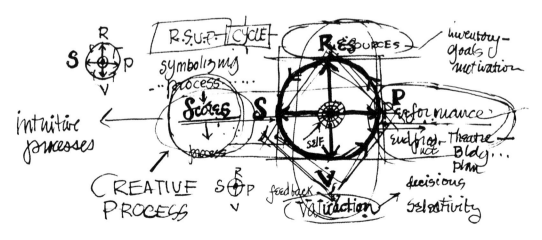

Figure 21-23. Urban planner Lawrence Halprin drew this sketch on a napkin to explain an idea about creativity to his daughter. As with many abstract idea sketches, this diagram requires verbal elaboration to communicate: read Halprin's *The RSVP Cycles: Creative Processes in the Human Environment,* published by George Braziller.

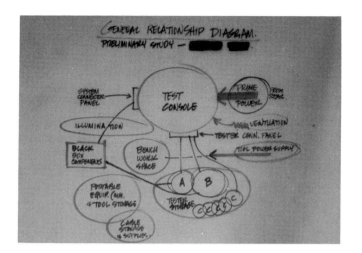

Figure 21-24. Designer Tony Chan drew this bubble diagram to view his design problem as a whole. Having abstracted basic relationships in the problem, he is now prepared to generate concrete solutions. His first solutions may not be correct, however; they may, in fact, reveal inadequacies in this initial bubble diagram. Repeated iteration from abstract to concrete, and back and forth again, will likely be necessary to solve the problem.

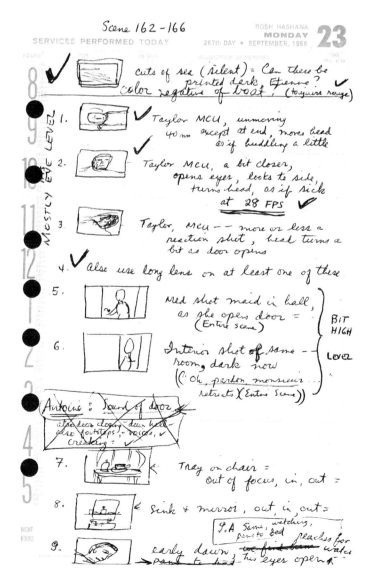

Figure 21-25. Movie writer and director James Salter sketches to decide the content, composition, and action of each camera shot, and to organize imagery into a dramatic sequence. These sparse notations, subsequently developed into a script, took final form in the film *Three*, produced by United Artists.

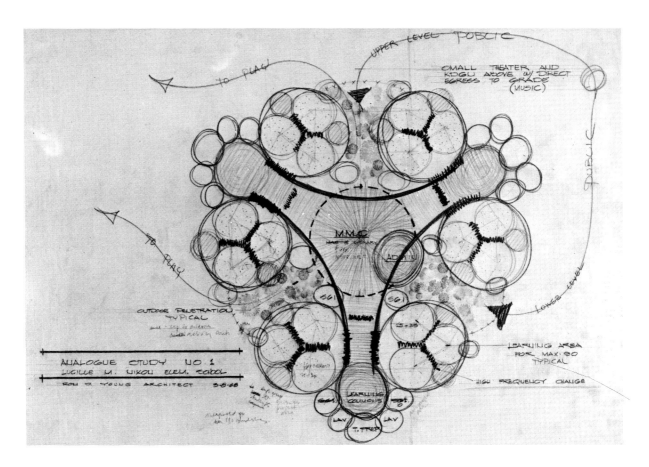

Figure 21-26. By its abstractness, this "analogue study" enabled architect Ron Young and his clients to agree upon fundamental functions and relationships to be provided by a new elementary school—without distraction by lesser details. (The sketch was drawn during a client conference.) Unlike the bubbles in Figure 21-24, these are spatially located. Young goes directly from this study to a spatially analogous, but more concrete, floorplan.

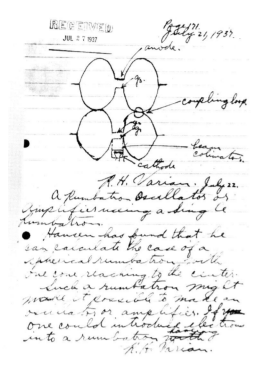

Figure 21-27. Inventor Russell Varian captured his initial idea for the rumbatron oscillator, or "klystron," in this simple schematic. You could not build Varian's design from this sketch: it is too abstract. On the other hand, Varian's notation embodies specific spatial features and relationships: it is therefore more concrete than Ligeti's musical notation, Chan's bubble diagram, or Halprin's visualization of the creative process.

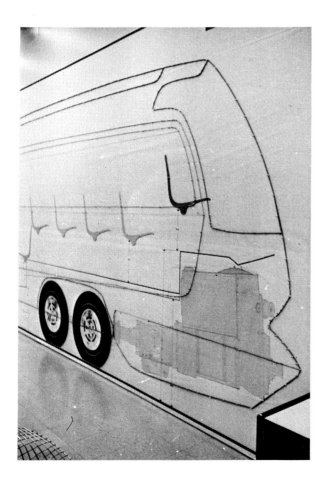

Figure 21-28. This full-scale schematic of a bus concept consists of shapes of colored paper (for the seats and engine) and lines of colored yarn—all held in place by large pull tacks. This mode of representation, developed by General Motors designers, is a boon to thinking because it is so easily manipulated.

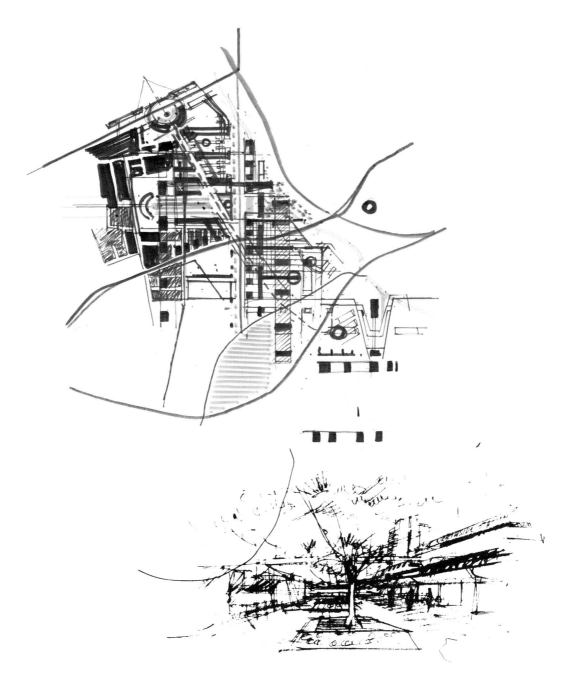

Figure 21-29. To consider an urban transportation concept, designer Gil Born first abstracted the idea as a line of movement through the city (drawn in red in the original). From this abstract "bird's-eye view," he then turned to a more concrete "pedestrian's-eye view" of the same concept (seen as an elevated train along a landscaped shopping mall). Such flexible use of abstract and concrete graphic language exemplifies the basic theme of this chapter.

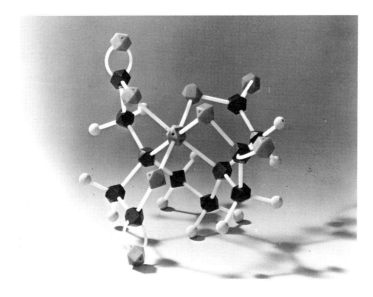

Figure 21-30. This Benjamin/Maruzen snap-together model of a calcium EDTA molecule is an example of a three-dimensional chemical diagram. Organic chemists who must deal with extremely complex molecular structures find such models invaluable tools for thinking and an essential alternative to two-dimensional diagraming.
Reproduced by permission of W. A. Benjamin, Inc., distributor of the Benjamin/Maruzen HGS Molecular Model Kits.

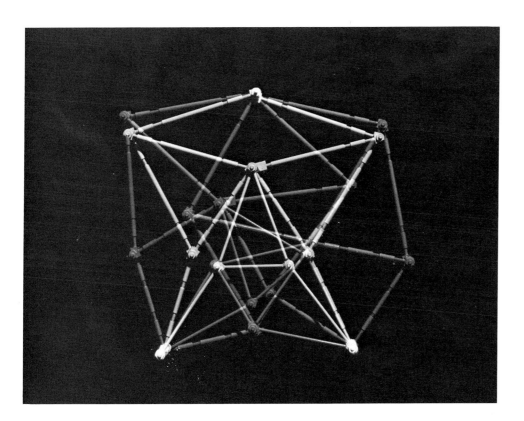

Figure 21-31. This dynamic three-dimensional diagram, constructed by a Stanford design student using the Swedish Mark Sylwan FAC X2 Construction Kit, actively demonstrates a basic theorem in mechanism kinematics. Almost as abstract as a line drawing, this model shows the action of this ten-link mechanism as no drawing could.

Concrete Graphic Languages

The graphic languages of orthographic, isometric, oblique, and perspective projection are usually directed toward the description of an actual thing. Orthographic and perspective projection are described in Chapter 13; isometric and oblique projection, essentially perspective shortcuts, are shown in Figure 21-32.

An additional concrete way to represent ideas is by means of the *three-dimensional model* or *space sketch* shown in Figure 21-33. Although not strictly a "language," three-dimensional modeling fits naturally at the end of the abstract-to-concrete continuum of graphic languages.

The next few pages contain examples of idea-sketching intended to show how concrete graphic languages and three-dimensional modeling can be elaborated. Notice, for example, how perspective drawings can be used to reveal structure by cross-sectioning, peeling away, exploding, and making transparent.

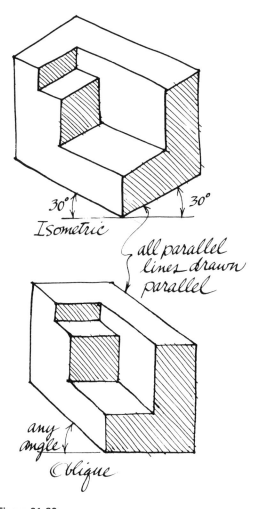

Figure 21-32.

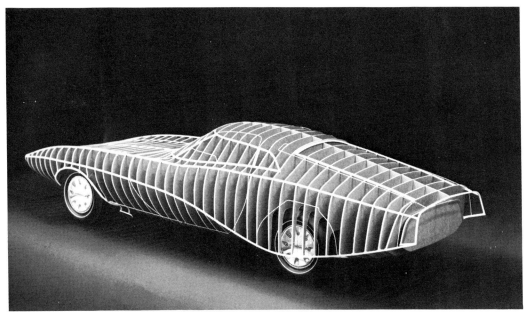

Figure 21-33.

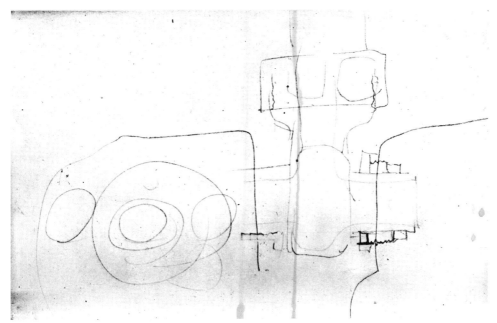

Courtesy of the Ford Archives, Henry Ford Museum, Dearborn, Michigan.

Figure 21-34. This drawing by Henry Ford appears to be abstract, but the screw thread on the right suggests that it depicts a concrete idea. Found in one of Ford's personal notebooks, it was not intended to communicate, and it does not. The staff at the Ford Archives in Dearborn cannot identify what it represents.

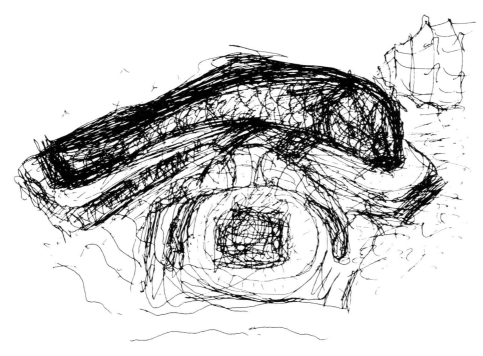

Figure 21-35. The demarcation between abstract and concrete graphic languages is of course arbitrary. This conceptual drawing for the *Grotto for Meditation* by architect Frederick Kiesler also seems, at first glance, to be quite abstract. However, notice that Kiesler's sketch resembles Figure 9-2 and exercise 9-3, Exploring the Object. Kiesler's marker is likely exploring a concrete idea.

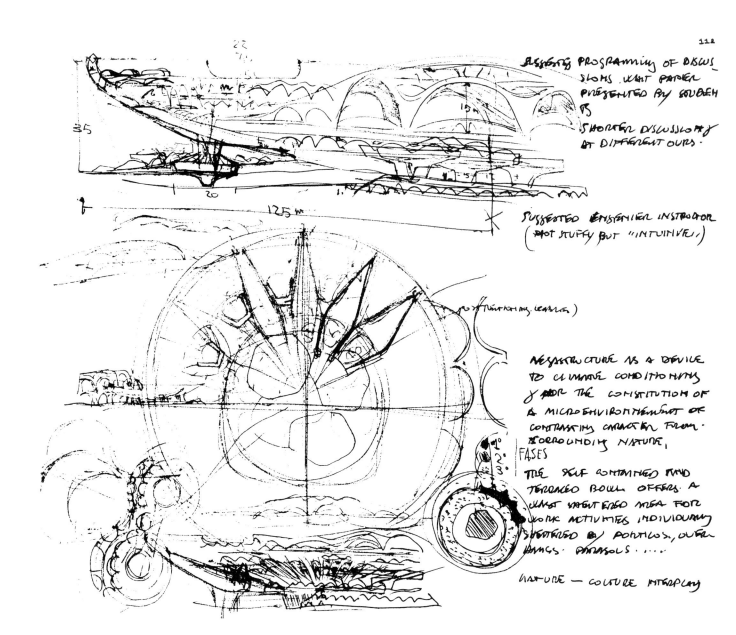

Figure 21-36. Although also loosely drawn, this sketch by visionary architect Paolo Soleri is a concrete graphic statement: it utilizes orthographic projection, it is dimensioned, it is concerned with a specific configuration. Observe the rhythm and grace with which Soleri conveys his ideas into sketch form.

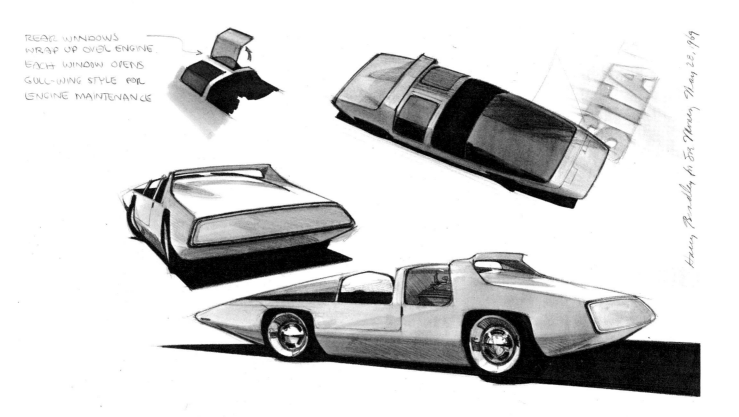

REAR WINDOWS
WRAP UP OVER ENGINE
EACH WINDOW OPENS
GULL-WING STYLE FOR
ENGINE MAINTENANCE

Figure 21-37.   A dramatic use of concrete graphic language occurs when one explores the appearance of an object of the mind without having to build it. Designer Harry Bradley examines a car concept from several angles this way, using one black and two gray felt-tip markers and a black Prismacolor pencil to portray its sculptural form.

Figure 21-38. Fashion designer Joyce Bradley predicts the appearance of a dress idea from two viewpoints, with construction notations alongside. Her choice of sketching media is a Rapidograph pen and colored felt-tip markers. Notice that she attaches a sample of actual dress materials to her drawing.

Figure 21-39. Film writer and director Federico Fellini began his professional life as a cartoonist. This makeup and costume sketch for the "white clown" in *The Clowns* typifies the idea-sketching that Fellini does for all his films. "Any ideas I have," he once said, "immediately become concrete in sketches."

Figure 21-40. This page from the notebook of designer Clair Samhammer records part of his imaginative response to an intriguing assignment. Asked to reconfigure the classic fig newton, Clair used his ability to sketch rapidly in perspective to "take snapshots" of the many mental images that emerged in his consciousness as he thought about the problem. Graphic ideation is a kind of camera for the mind's eye.

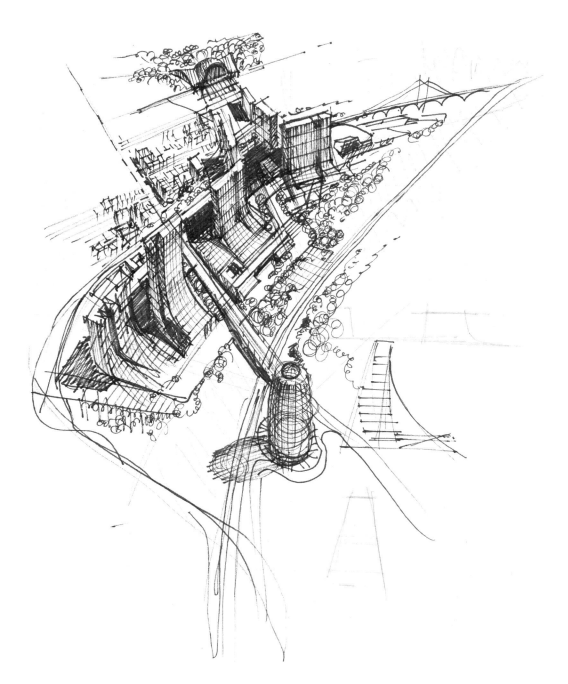

Figure 21-41. No camera, no airplane, and no buildings:
Gil Born's thumbnail sketch is a snapshot of his predictive
imagination. The value of making ideas visible on paper
before building them is especially evident with projects of
this size.

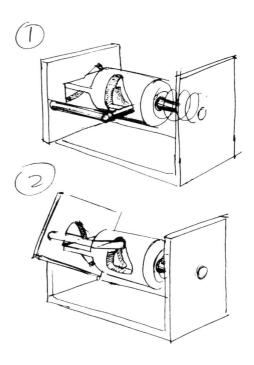

Figure 21-42. Designer Del Coates is not concerned primarily with appearance; he has chosen perspective for its ability to help him visualize this mechanism's spatial action. A single perspective view is superior to the flat, multiple images of orthographic projection for this kind of mental operation. Orthographic views will be used later to consider proportions and dimensions.

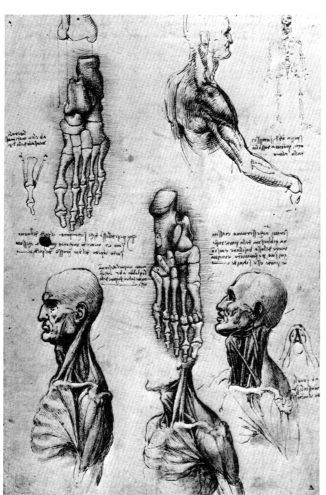

Figure 21-43. Thinking about external surfaces without also thinking about the internal structure is, by definition, superficial. The dissective capability of concrete graphic languages is illustrated in this drawing by Leonardo da Vinci

Royal Collection, Windsor Castle, copyright reserved.

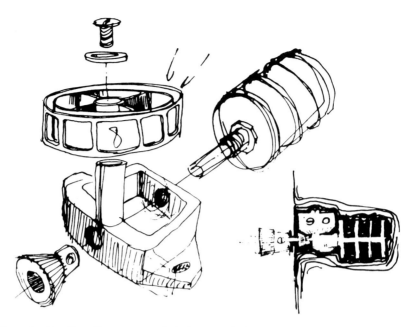

Figure 21-44. Tony Chan uses "exploded perspective" to visualize how the parts of this mechanism will be assembled. His ability to draw freehand enables him to clarify his thinking on this problem quickly.

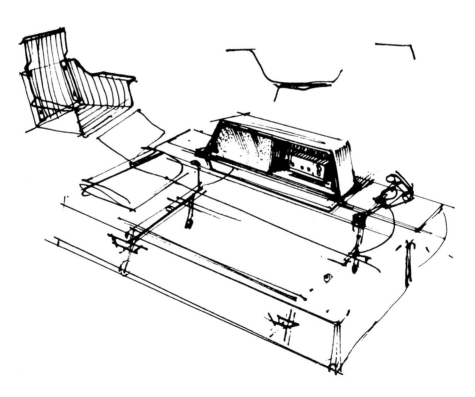

Figure 21-45. Gil Born uses perspective to make the structure of his furniture concept transparent. Cross-sectioning, peeling away, exploding, and transparentizing enable the visual thinker to explore a more complete reality than is portrayed by so-called "realistic" sketches of external appearance.

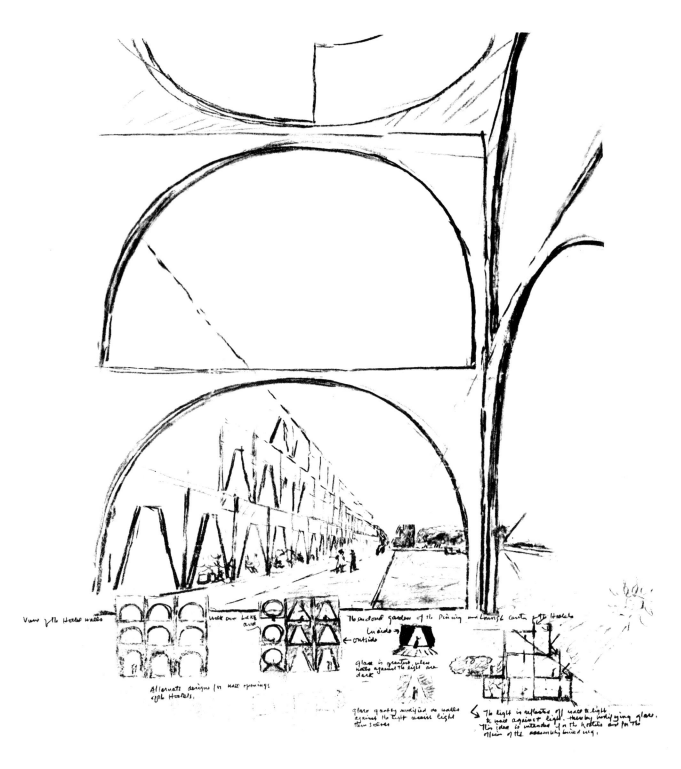

Figure 21-46. Architect Louis Kahn uses language—
verbal notation, orthographic projection, and
perspective—to probe his thinking from varied viewpoints
and levels of abstraction.

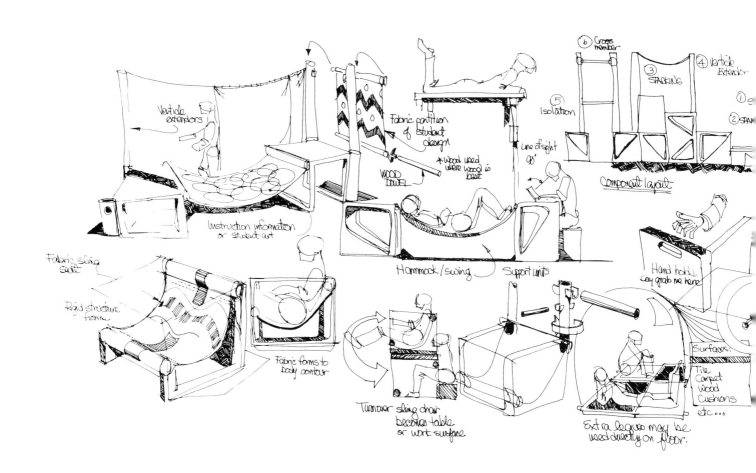

Figure 21-47. Another avenue to reality is exploring one's understanding of a problem by generating alternative solutions to it. In this scroll of idea sketches, designer Brooks Stover is conceptualizing school furniture that will enable children to create their own learning environments. Each sketch represents a look at this problem from a different viewpoint.

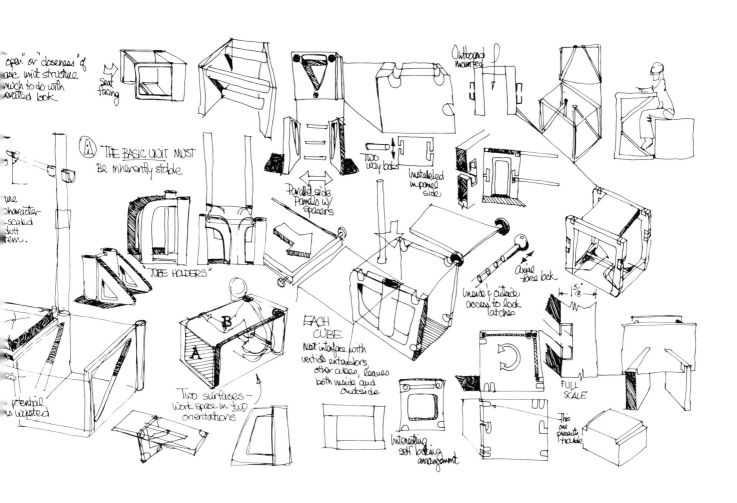

"open" or "closeness" of our unit structure much to do with revealed look

seat facing

Outboard mounted

Ⓐ THE BASIC UNIT MUST Be inherently stable

are character scaled Jutt em.

Two way bolts

Installed in panel side

Parallel side panels w/ spacers

axial force lock

inside & outside access to lock latches

1⁵/₈"

"TUBE HOLDERS"

EACH CUBE

Must interface with vertical extenders, other cubes, leaves both inside and outside

FULL SCALE

potential is wasted

A B

Two surfaces – Work space in two orientations

Interesting self locking arrangement

↻

This one presents trouble

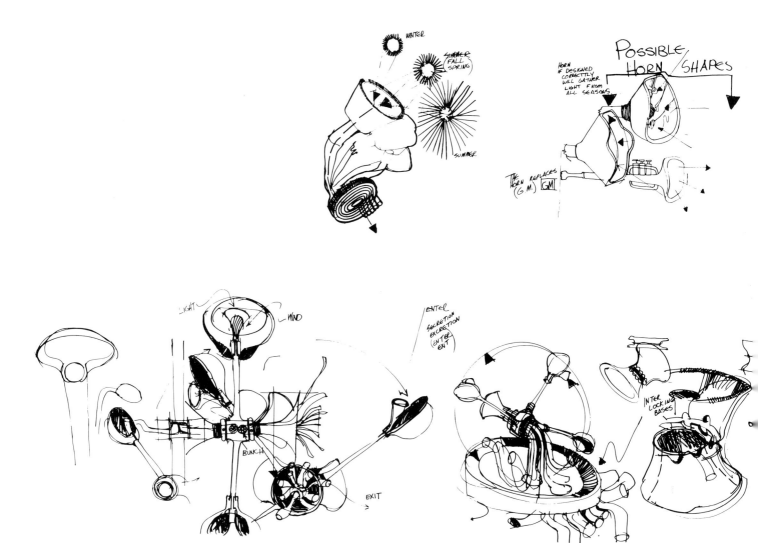

Figure 21-48. Drawing on long scrolls seems to encourage copious idea-sketching; perhaps the thought-stream flows more readily onto endless paper than onto single sheets. In this exceptionally fluid scroll, designer Ed Wittner is exploring ideas for a sun clock.

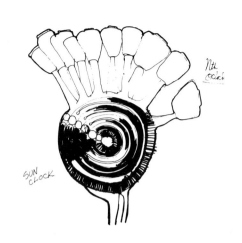

SUN
CLOCK

FINGERS
LIGHT
PLAYING ON
THE CRYSTAL
SURFACE

LIKE CABLES
OF WIRE
SO CABLES
OF ACRYLIC
PLASTIC
WILL NOT
SNAP
IN THE
WIND !!!

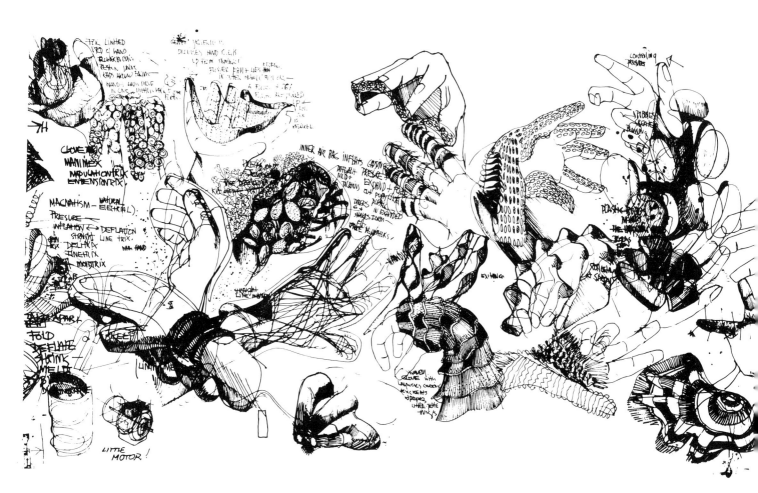

Figure 21-49. Designer Nancy Strube's scroll records her Synectics-like excursion (see Chapter 18) in search of a glove with special gripping powers. The sketches shown are based largely on analogies to biological forms that grip. In the verbal notations between sketches, Nancy engages in word-play (see "relabeling," Chapter 8) to recenter her perceptions of the problem.

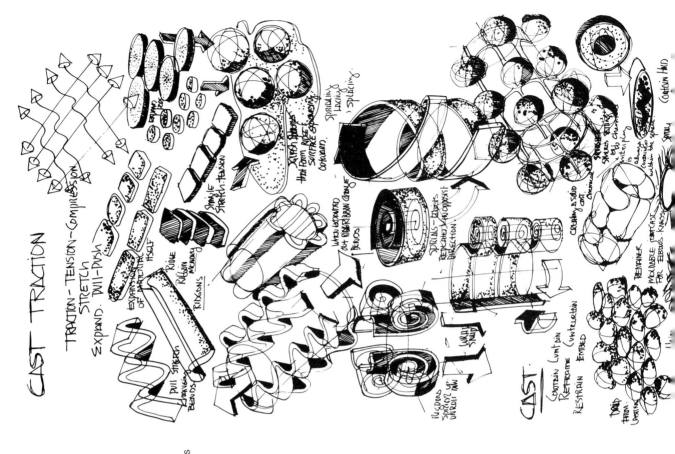

Figure 21-50. This vertical scroll, also by Nancy Strube, records her ideas for a new casting material and process. Compared with her previous drawings, these idea sketches are more abstract.

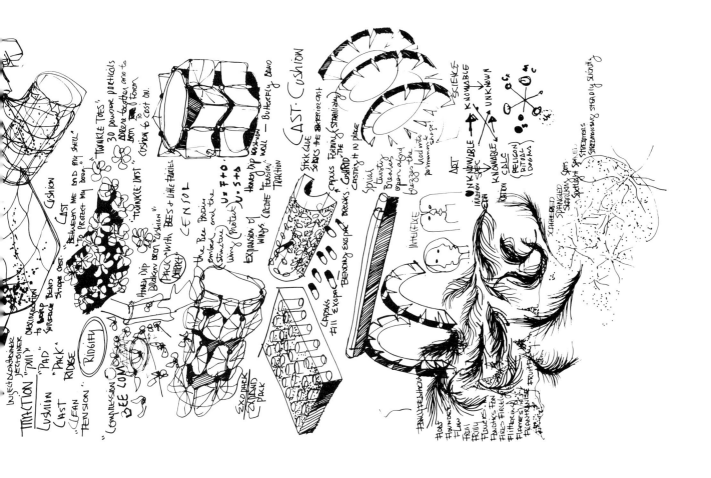

Figure 21-51. In a technique he calls Generative Graphics, architect Joseph Brunon uses large, wall-mounted scrolls to facilitate group problem-solving. As ideas are generated by a group, Brunon records them on the scroll with a cartoon, diagram, or written phrase, using large, colored felt-tip markers. He does not problem-solve himself, but works to represent ideas fairly, to stimulate full and energetic participation, and to find graphic ways to illustrate patterns that evolve in the group's thinking.

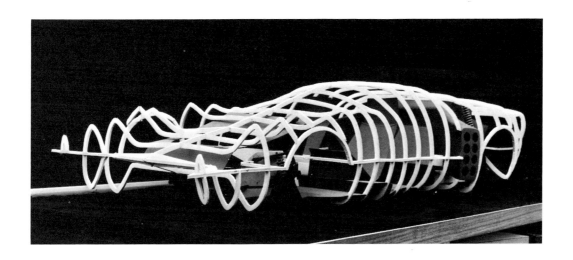

Figure 21-52. As visual thinking becomes increasingly concrete, it usually takes three-dimensional form. This model by designer Mike Golden still contains considerable abstraction: the outer shell of the car is symbolized by white lines in space, the seat by a bent plane, the engine by a solid block, and so on. Interior components are also brightly color-coded to denote their function.

Figure 21-53. Designers of the proposed National Data Buoy used this "soft mockup" to determine optimum spatial arrangement of mechanical, electrical, electronic, and fluid subsystems within a compact hull. Soft mockup materials such as Foamcore facilitate rapid and inexpensive spatial organization and easy modification. Drawing is the difficult way to solve problems of this order.

Figure 21-54. Full-scale mockups were used extensively to develop and evaluate ways to transport the Lunar Rover vehicle on the Apollo 15 flight to the moon. This soft mockup revealed the feasibility of an early piggy-back version. The interior of this mockup was also detailed to permit designers to study the human-factor feasibility of the module's instrumentation and controls. Easily constructed mockups help to relate group efforts and to focus thinking on mutual problems. Three-dimensional sketching should not be left to the last; it is often useful at early stages of thinking (see Chapter 7, Externalized Thinking).

Figure 21-55. Le Corbusier's three-dimensional sketch for a church suggests that model-making precision is not necessary for productive visual thinking. Indeed, the very roughness of the sketch encourages modification, manipulation, and consequent evolution of the spatial concept.

Much as writers often organize and refine their thinking by writing several drafts, visual thinkers can manipulate and reform their ideas by drawing an evolving sequence of graphic images. Graphic images are easily developed in this way by using tracing paper (21-3).

21•3 MODIFY BY OVERLAY

Tracing paper is extremely useful for developing ideas. Overlay an initial sketch with tracing paper; then:

1. Modify its proportions.

2. Change its details.

3. Add, subtract, or reposition its elements.

4. Simplify it.

5. Examine its internal structure.

6. Manipulate its elements: using different-colored markers, draw separate elements on each of several pieces of *very transparent tracing paper,* and then stack and manipulate elements in relation to each other.

7. Experiment with color schemes.

Whatever graphic language you use, add color (21-4). *Color brings excitement and clarity to idea-sketching, and it is reflected directly into excitement and clarity in thinking.*

21•4 CLARIFY WITH COLOR

Use color to:

1. Color-code similar functions (in a chart, in a schematic, in an orthographic cross-section).

2. Distinguish one part from another (especially useful in complex cross-sections).

3. Clarify elements that are confusedly overlapped in space (felt-tip markers provide an excellent effect of transparency).

4. Describe motion (such as traffic flow in a flow plan, or alternative positions of a moving part).

5. Dramatize critical areas.

6. Strengthen, by color contrast, a figure-ground relationship (for example, place a patch of color "behind" a line drawing).

7. Define, by shading, an otherwise ambiguous characteristic of form.

To obtain graphic-language flexibility, practice moving an idea through the varied viewpoints and mental operations of several graphic languages, as in exercise 21-5.

21•5 ABSTRACT-TO-CONCRETE

Using both abstract and concrete graphic languages, see how many ways you can think about a single theme. Take, for example, yourself. Here's a starter: a family tree, or a diagram showing your relationships within your immediate circle of friends. How many other graphic means can you devise to express yourself?

Notice, in the previous exercise, how changing to another graphic language automatically changes your *viewpoint*. The concrete language of orthographic projection, for example, brings your viewpoint "close in"; a more abstract diagram allows you to step back and view basic relationships. Notice also that each language automatically sets your thinking into certain *mental operations*. You cannot use orthographic projection without using the operation of rotation; you are also invited to take a cross-section; the head-on viewpoints of this language insist that you consider dimensioning and proportional relationships. In addition to the powerful operation of abstraction, a diagram allows you to organize events in the dimension of time, for example (as in a flow diagram), or to quickly reorder relationships between abstract ideas. Because of the viewpoints and operations built into each graphic language, moving from language to language is one of the easiest strategies for expanding the scope and extending the operational power of your thinking. Finally, *the most concrete graphic languages lead very naturally into externalized thinking,* as discussed in Chapter 7. The result of such three-dimensional thinking is often called a "space sketch" (21-6).

21•6 SPACE SKETCH

Using easily worked materials (as suggested in Chapter 7), form and manipulate your idea in three dimensions.

References

1. Vygotsky, L. *Thought and Language.* The M.I.T. Press.
2. Ogden, C., and Richards, I. *The Meaning of Meaning.* Harcourt Brace Jovanovich.
3. Bruner, J. *On Knowing: Essays for the Left Hand.* The Belknap Press of Harvard University Press.
4. Cassirer, E. *Language and Myth.* Dover.
5. Picasso, P. Quoted by Zervos, C., in *The Creative Process,* edited by B. Ghiselin. New American Library (Mentor Books).
6. Hill, E. *The Language of Drawing.* Prentice-Hall (Spectrum Books).
7. Neisser, U. *Cognitive Psychology.* Appleton-Century-Crofts.
8. Hayakawa, S. *Language in Thought and Action.* Harcourt Brace Jovanovich.
9. Alexander, C., et al. *A Pattern Language Which Generates Multi-Service Centers.* Center for Environmental Structure, 2701 Shasta Road, Berkeley, California.

THE STRATEGY APPROACH

22

Outwitting the Enemy

Webster's dictionary defines *stratagem* as "an artifice or trick in war for deceiving and outwitting the enemy." For *strategy* it gives us a gentler meaning: "a careful plan or method; the art of devising plans or stratagems toward a goal."

Although the second definition comes closer to the intentions of this book, the reference in the first to "outwitting the enemy" does ring a metaphorical bell. Faced with an intractable problem, the problem-solver can choose to retreat, to surrender, to negotiate an inadequate compromise—or to outwit the enemy. What follows is a discussion of artifices, tricks, and schemes for thinkers who fully intend to create satisfying solutions to seemingly unsolvable problems.

Strategies Are Tools

David Straus provides us with a peaceful analogy by comparing the acquisition and use of strategies to a carpenter and his tools:

A carpenter has a tackboard full of tools in front of him. Each one has a range of uses and limitations. The carpenter has used them all in a variety of situations. When he bought a new spokeshave, it took him awhile to get used to it, to get the feel of it, and he did this by experimenting with it on many different kinds of woodworking problems. When he bought a new lathe, certain kinds of problems became easier for him. Problems that could be solved by round and symmetrical forms became simpler, [and] the lathe permitted him to do things he couldn't do before. Once he has mastered the use of a tool, it becomes almost an extension of his hand. When a piece of wood is too rough, he reaches for the plane or sandpaper without consciously stopping to think about it. Given most problem situations, he can quickly decide what kinds of tools might be used to resolve [them], and what the probable effect of each tool will be. In many situations he knows there is no one right tool, but he . . . is more comfortable with one over another. . . . His knowledge and skill with his tools . . . determine a substantial part of his overall ability as a carpenter.[1]

To take the analogy further, the kind of tool that you would want to keep in your toolbox at all times is the kind that has many applications. A screwdriver, for example, can be used to repair a sewing machine, to adjust a carburetor, to assemble a clock. Similarly, the most effective strategies are the ones that can be used in a wide variety of applications. The strategies we will discuss here can be used in scientific thought, artistic thought, mathematical thought—indeed, in virtually every thinking situation. To be sure, as screwdrivers are sometimes used as hammers, chisels, or prybars by the ill-equipped or inept, strategies can also be applied inappropriately and ineffectively. As with tools, strategies must be skillfully chosen and applied in order for them to be effective.

A carpenter does not, of course, become skillful in using tools by merely looking through a hardware catalog. Similarly, you will not increase your repertoire of visual-thinking strategies by merely reading this book; visual-thinking strategies are not "facts" that can be acquired passively. Nor are you likely to expand your repertoire of thinking strategies until you feel the need to do so. *The way to learn about visual-thinking strategies is to use them actively and repeatedly to help you solve problems that interest you.*

Identifying a Challenge

People confronted with an unsolved problem that they find fascinating and worthwhile to solve stand a far better chance to develop their thinking abilities than people presented with a problem that bores them. Of course, a meaningful challenge to one person may very well prove to be a bore to another. An architect and an engineer will not likely be challenged by the same problem; indeed, two people working in the same discipline will not be equally challenged by the same problem. Only *you* can identify the kind of challenge that will stimulate you to think deeply.

If your profession actively involves you in solving challenging problems, it will probably be easy for you to choose a problem to work on. If not, you may want to try developing a list of problems that interest you (22-1).

22•1 CHALLENGE LIST

Keep a list of problems that you find to be challenging. If you read a magazine article about a problem that interests you, put it on the list. Add a "bug list" (things that bug you). As you proceed in this book, try to apply visual thinking strategies toward the solution of these problems that personally interest you.

Selecting a Strategy

Once you have chosen a problem, how should you select a strategy, or strategies, to solve it? One approach to strategy selection is essentially chronological. Just as a carpenter usually saws a log before sanding it, so a thinker usually applies thinking tools in some logically ordered sequence. For example, the sequence Express/Test/

Cycle (described in Chapter 20) is a chronologically organized approach to graphic ideation. An important value of the sequential order of ETC is the way it separates the expression of ideas (E) from subsequent judgmental testing (T) and clarifies the subsequent choice to cycle (C) to an altogether different thinking strategy.

A somewhat more elaborate chronology is shown in the Strategy Flow Chart that appears following this chapter. As the Flow Chart shows, basic thinking strategies have a time-oriented flow that can be diverted by choices at several junctures in the thought process. Following the Flow Chart, you will find a Strategy Index that defines the strategies depicted in the Flow Chart and refers you to skills sequences in the book that embody each strategy.

The purpose of the Strategy Flow Chart is to clarify that strategic choices always exist in problem-solving, whether you are aware of making them or not. *The essence of the strategy approach to problem-solving is making conscious choices from a fully perceived range of alternatives, rather than merely following habitual patterns.* For example, one choice that always exists is to reject the problem. Not every problem that looms into view is worth trying to solve; choosing whether or not to solve a problem can be an effective strategy for keeping your energy centered on important issues.

Turn to the Strategy Flow Chart on page 194 and follow along with a brief description of each of the strategies on the chart. Following the preliminary definition of a problem, the first strategy is to *relax and clear your mind.*

Relax/Clear

The Relax/Clear strategy should, in fact, be used more frequently than its single appearance on the diagram suggests. Chapter 6 describes why a state of relaxed attention is conducive to thinking generally, and why the specific relaxed state in which the conscious mind is *cleared of the problem* is especially valuable. Thinkers who appreciate the value of this strategy will consciously choose to divert their minds from the problem from time to time by taking a walk, for example, or even by sleeping on the problem. The Relax/Clear strategy acknowledges that *insights originating below the threshold of consciousness are more readily recognized by a mind that is calm and quiet.*

The particular placement of the Relax/Clear strategy in the Flow Chart is related to the next choice. At the critical juncture where you decide whether to stop thinking about the problem, to keep thinking about it in the same way, or to try a new tack, your mind must necessarily be clear. This specific choice obviously cannot be made if you are thinking about the problem all the while.

So relax, clear your mind, and confront the fundamental choice in all problem-solving.

Stop/Persist/Cycle

The fundamental choice in all problem-solving, often passed over without the slightest recognition, is:

1. Choose to *stop* (that is, refuse to accept the problem, or refuse to continue to think about it).

2. Choose to *persist,* if you have been thinking about the problem, in the use of your current thinking strategy.

3. Choose to *cycle* your thinking into another problem-solving strategy.

Stop? "I'd soon be fired if I chose not to solve the problems assigned to me!" Even the thinker whose concern for job security is paramount can afford to pause for a moment to consider the power of the choice to accept the problem or not. Perhaps the problem is *not* valid, either for you or for your employer. *A brilliant solution to some problems is simply to* dis-*solve them!* Or perhaps you are sitting on the fence, neither accepting nor rejecting the problem. Clearly, one of the most devious enemies of effective problem-solving is

half-hearted engagement with the problem. Don Koberg and Jim Bagnall, in their excellent guide to creativity, *The Universal Traveler,* call the first step in all problem-solving "acceptance."[2] Only with enthusiastic acceptance of the problem can your best thinking energies come forth. The stuck place between "stop" and "go" is experienced as frustration and boredom. So relax, clear your mind, and *choose!*

Persist or cycle? If you choose to accept the problem, the Flow Chart offers a second choice: should you persist with your current strategy or opt to cycle your thinking into an alternative strategy? When should you abandon one strategy to try another? When is perseverance a virtue, and when is flexibility? Sometimes dogged persistence in the use of a single strategy yields a solution: despite frustration and fatigue, the thinker rattles the same key in the door until it finally opens. On the other hand, it may simply be the wrong key. When staying with one strategy pays off, we call it "perseverance"; when it does not, we call it "stubborn inflexibility." Genius is often associated with the ability to persevere or, as Edison said, to perspire. Creativity is also linked to the ability to be flexible. Clearly, we are facing a paradox. Perseverance and flexibility are opposites that together form an important unity.

Flexibility is especially important at the onset of problem-solving. Flexibility permits recentering, thereby diminishing stereotyped ways of seeing the problem and increasing the opportunity for generating fresh solutions. However, flexibility can also be abused: the thinker who flits from one strategy to another, never using any strategy in depth, is comparable to the impatient carpenter who uses one tool after another to make a simple cut. As long as one tool, or one strategy, is clearly working, changing to another (simply to be flexible) is ill-advised. Perseverance is in order whenever a thinking strategy is yielding new viewpoints, new information, or additional alternatives in the exploratory stages of problem-solving. In developmental stages of thinking, perseverance is especially important. *Once the basic idea has been discovered, often by exploratory flexibility, the hard work of giving that idea coherent form usually requires tremendous perseverance.*

The proper balance of perseverance and flexibility becomes more evident with experience. Much as a carpenter learns by experience when to abandon one tool for another, the thinker learns when to cycle his thinking by means of another strategy, and when it is more appropriate to continue with the present one.

Left or Right Hemisphere?

The next choice in the chronologically ordered flow of strategic choices presents the problem-solver with the challenge to balance preference for thinking in one hemispheric mode by consciously directing thinking to the other hemispheric mode. (See Chapter 4, Ambidextrous Thinking, for a review of the rationale for this strategy.) Because this is a book of essentially right-hemisphere strategies, the reader who seeks strategies for the left hemisphere is referred to Recommended Reading at the end of this chapter for source books. Problem definition is an example of an essentially verbal, left-brained strategy. The logical, objective, language-oriented operations of the left brain are also invaluable for evaluating solutions; for gathering information about the problem; for measuring, systematizing, and otherwise imparting discipline and order to the problem-solving process; and of course, for communicating.

Right-brained or visual-thinking strategies are diagramed on the page following the Flow Chart. As shown, each of the six basic strategies (Abstract, Concretize, Modify, Manipulate, Transform, and Timescan) can be applied in the context of seeing, imagining, or drawing. Also, each strategy flows directly into the Express strategy, and then into the Test strategy, as described in Chapter 20. As the diagram indicates, there is no step-by-step methodology for choosing a visual-thinking strategy. Instead, your choice must be based on your knowledge

of what each strategy can do—which brings us to a discussion of the Strategy Index.

Using the Strategy Index

Turn now to the Strategy Index that begins on page 193. *This section transforms the book into a strategy manual for problem-solving* by using the book's exercises to illustrate specific thinking strategies.

The Strategy Index begins with the strategies on the Flow Chart and proceeds to the visual-thinking strategies. It gives a brief description of every strategy and refers you to skills sequences in the book that demonstrate how the strategy can be applied. When you refer to a skills sequence, you will sometimes need to look to its *underlying principle* to see the strategy. For example, 7-2, Tangrams, is listed under the strategy Manipulate in the context of seeing: the underlying principle here is to represent your problem with paper cut-outs that you can easily shift around into a variety of configurations. In another instance, 3-8, Rotating Dice, is listed under Manipulate in the context of imagining: here the underlying principle is to rotate the object of your thinking in your mind's eye. Every reference contains an underlying principle that, with a little imagination, you can apply directly to your problem.

When the choice of a strategy is no longer part of an ordered, chronological flow, on what basis do you choose? Here are three guiding principles:

1. *Contrast.* Seek a strategy that contrasts sharply with the one you have been using. If you have been thinking with inner imagery, turn to idea-sketching; if you have been thinking abstractly, move your thinking toward the concrete. By changing to a contrasting vehicle and/or mental operation, you will automatically expand your thinking into new territory.

2. *Function.* Seek a strategy that performs a necessary function. If you have just defined the problem, a logical next step

is to generate alternatives; if you have developed an idea sketch that looks promising, an excellent next step may be to build a model. By choosing a strategy that performs a needed function, you are using the strategy as a tool to develop and clarify your thinking.

3. *Intuition.* Scan the available strategies and choose the one that feels right. By being willing to play a hunch, you allow your thinking to move along subconscious paths to choose an appropriate strategy. In fact, all choices contain an element of intuition, and every choice is best made in the relaxed, clear state of mind that allows intuitive insights to emerge into consciousness and be recognized.

Once you have chosen a strategy, the Express strategy automatically joins forces to give it visible form: no strategy is complete until it is expressed. Then you apply the Test strategy and choose whether to accept the solution or to return to another cycle of problem-solving.

Full Circle

In Chapter 1 I suggested that a primary purpose of this book is to encourage flexibility in levels, vehicles, and operations of thinking. Now let us review how you have done. If you have experienced thinking below the level of waking consciousness in hypnogogic reveries and dreams and have also consciously expressed your thinking in the form of graphic language, you are certainly obtaining flexibility in levels of thinking. If you have experienced the three visual vehicles of thought and have related these to your other senses and to the symbolic vehicles of verbal or mathematical language, you are clearly acquiring flexibility in thinking vehicles. And as this last chapter has suggested, if you have actively tried most of the experiences in this book, you are also attaining flexibility in thinking operations.

I hope that you are also experiencing the *value* of thinking flexibility. As suggested in Chapter 4 and elsewhere, the most profound sort of creative thinking requires transfer between unconscious and conscious levels of mental activity. However, reading about this is one thing and experiencing it is another. You will not be convinced until you experience it, perhaps by gaining a valuable insight in a dream. Similarly, you have read that flexibility in thinking vehicles leads to flexibility in thinking operations—that visual thinking, for example, encourages thinking operations not possible or not readily performed in the verbal mode, and vice versa. Again, the value of such flexibility will escape you until you experience it. *To experience the value of flexibility, you must persevere to the point where limitations are clearly experienced in one mode of thought and benefits clearly obtained by moving to another.* In each section, I have discussed the limitations as well as the benefits of thinking by seeing, imagining, and drawing. You will experience the value of flexibility when you experience how seeing advances your thinking to a point, then how imagining frees it to advance further, then how idea-sketching overcomes limits inherent in imagining, and so on.

We have come full circle. Since seeing, imagining, and idea-sketching are endlessly related, this could as well be the beginning as the end of the book. I hope that you will continue to use it, moving back and forth between experiences, obtaining greater competence in thinking skills. And I hope that these experiences will challenge you to invent new ones. When you find another useful way to educate visual thinking, would you please write to me about it, in care of the publisher? Books must end, even when curiosity grows.

References

1. Straus, D. In *Tools for Change: A Sourcebook in Problem-Solving.* Interaction Associates, Berkeley.
2. Koberg, D., and Bagnall, J. *The Universal Traveler.* William Kaufmann.

Recommended Reading

The following books describe left-hemisphere problem-solving strategies that complement the essentially right-hemisphere strategies listed in the Strategy Index:

Allen, M. *Morphological Creativity.* Prentice-Hall.
Flesch, R. *The Art of Clear Thinking.* Harper.
Hayakawa, S. *Language in Thought and Action.* Harcourt Brace Jovanovich.
Kepner, C., & Tregoe, B. *The Rational Manager.* McGraw-Hill.

Nevell, A., & Simon, H. *Human Problem Solving.* Prentice-Hall.
Osborn, A. *Applied Imagination.* Scribner's.
Prince, G. *The Practice of Creativity.* Collier Books.

See Parnes, S., Noller, R., & Biondi, A., *Guide to Creative Action* (Scribner's) for an extensive problem-solving bibliography, a listing of some 175 films on creativity, and 100 problems and exercises for creative thinking.

For an excellent catalog of books and classroom materials that relate visual thinking to mathematics, write to Creative Publications, P.O. Box 10328, Palo Alto, California 94303.

STRATEGY
INDEX

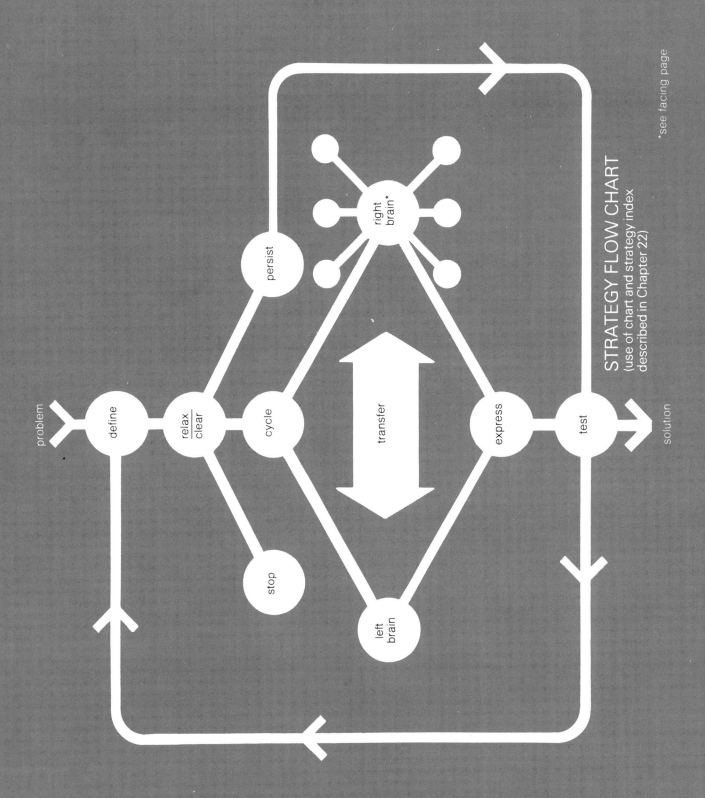

STRATEGY FLOW CHART
(use of chart and strategy index
described in Chapter 22)

*see facing page

problem

define

relax / clear

cycle

stop

persist

right brain*

transfer

left brain

express

test

solution

VISUAL-THINKING STRATEGIES

DEFINE

Recognize problem
Question/assess requirements/identify constraints
Gather information
State problem in writing
State problem graphically
List objectives/criteria

Refer to:
19-2   Envision a Goal
20-8   Criteria Formulation
21-2   Pattern Language

RELAX/CLEAR

Relax
Divert conscious attention from problem
Mediate/incubate
Clear/open mind to intuitive insights

Refer to:
Chapter 6   Relaxed Attention
6-12   Clearing the Ground
16-2   Lazin' Down the River
19-4   Incubate
19-5   Leaping

STOP

Reject problem/dis-solve it

Refer to:
17-4   Stop!
17-5   Flood
6-12   Clearing the Ground

PERSIST

Continue thinking with present strategy
Increase/decrease effort

Refer to:
6-8   Attention Is Undivided
6-9   Attention Follows Interest
6-10   Attention Is Dynamic
6-11   Attention Is Continual Aha!

CYCLE

Choose another strategy

Refer to:
20-5   Recenter
20-9   Cycling
Chapter 8   Recentering

LEFT OR RIGHT HEMISPHERE?

Choose one hemispheric mode
Internal transfer
Ambidextrous thinking

Refer to:
4-1   Food for Thought
4-3   Internal Transfer

ABSTRACT

Defocus/withdraw from details/generalize
Look at the big picture/seek overall pattern
Induce principles from particulars
Think with broad brush
Group/classify/symbolize
Diagram/chart/schematize
Analyze/repattern
Expand viewpoint/think divergently
View problem as part of a system/see problem's context

Refer to:

See
3-4   Categorizing
3-14   Visual Induction I
10-2   Grouping
10-4   Focus!
10-5   Da Vinci's Device

Imagine
14-4   Phosphene Projections
18-4   Abstract Word-Images
18-5   Synectics Excursion

Draw
10-1   Fingerpaint Patterns
21-2   Pattern Language
21-5   Abstract-to-Concrete
(see also illustrations, Chapter 21)

MODIFY

Reform elements of problem or solution
Exaggerate/embolden/distort
Modify slightly/understate/refine
Reproportion: e.g., heighten, thicken, deepen
Clarify
Modify nonvisual quality: e.g., soften/lighten
Unify/standardize/modularize

Refer to:

See
12-5   Subtle Changes

Imagine
18-2   Ratio

Draw
12-6   Caricature
21-3   Modify by Overlay
21-4   Clarify with Color

TRANSFORM

Transform identity of problem or solution
Abandon labels, conventions, stereotypes
Think metaphorically/seek analogies
Empathize/role play

Refer to:

See
3-13   Spatial Analogy
8-2   Topsy-Turvies
8-3   "Rediture"
8-4   A Rose Is a Cork
8-5   Visual Similies
10-5   Da Vinci's Device

Imagine
14-6   Control of Mental Imagery
16-1   Hypnogogic Imagery
16-4   Productive Dreaming
17-6   Worry-in-Reverse
17-8   Merged
18-5   Synectics Excursion

Draw
9-4   Expressive Line
20-1   Thirty Circles

MANIPULATE

Rearrange elements of problem or solution
Disassemble/separate/reassemble
Randomize/play with
Eliminate/add on
Substitute/combine
Transpose/superimpose
Reverse/rotate/unfold
Cross-section/transparentize
Organize/systematize

Refer to:

See
7-2   Tangrams (and Chapter 7)
8-2   Topsy-Turvies
21-6   Space Sketch

Imagine
2-4   Painted Cube
3-9   From Another Viewpoint
3-10   Folded Pattern
7-3   Soma Cube
14-6   Control of Mental Imagery
17-1   Breathing
17-7   Object-in-Space
18-1   X-Ray Vision
18-3   Changing Places

Draw
13-3   Cross-Section
21-2   Pattern Language
21-3   Modify by Overlay

CONCRETIZE

Converge thinking/focus in on one solution
Clarify/realize/actualize/exemplify
Develop details
Use concrete graphic languages
Embody concept in 3D

Refer to:
See
2-3 Spaghetti Cantilever
8-1 "Feeling the Actual"
9-3 Exploring the Object
21-6 Space Sketch

Imagine
14-5 Clarity of Mental Imagery
15-7 Apple II
17-7 Object-in-Space

Draw
13-3 Cross-Section
13-6 Two-Point Perspective
21-4 Clarify with Color
21-5 Abstract-to-Concrete

EXPRESS

Explore, develop, express ideas
Generate alternatives
Fluency-flexibility
Defer judgment/defy censorship/leap in/blue sky
Change viewpoints/alternate strategies
Ask "what if?"
Record

Refer to:
See
7-1 Tower of Pulp
7-2 Tangrams
21-6 Space Sketch

Imagine
16-2 Lazin' Down the River
16-3 Dream Diary
16-4 Productive Dreaming
18-5 Synectics Excursion
19-5 Leaping

Draw
10-3 Scope Thumbnails
20-1 Thirty Circles
20-2 Visual Brainstorming
20-3 Idea Log
20-9 Cycling

TIMESCAN

What from the past can help you now?/retrieve
What from the now can help you now?/observe
Envision goals/plan
Foresee consequences/predict

Refer to:
See
8-1 "Feeling the Actual"

Imagine
15-2 Childhood Home
19-2 Envision a Goal
19-3 Envision Consequences

Draw
2-5 With One Line
15-4 Memory Drawing
19-1 Look Ahead

TEST

Take critical viewpoint
Display/compare
Evaluate according to criteria
Discriminate pros and cons
Record your evaluation

Refer to:
See
2-2 Cards and Discards
3-3 Matching
11-1 Paint Chip Hunt

Imagine
19-3 Envision Consequences

Draw
20-4 Display Your Graphic
20-5 Memory
20-6 Recenter
20-6 Compare
20-7 Colored Notations
20-8 Criteria Formulation

ACKNOWLEDGMENTS

Although I originally intended this to be a small picture book of idea-sketches, on the way to the publisher I was intercepted by a number of persuasive people. What follows is my expression of thanks to those who influenced me to evolve my initial objective to its present form.

My greatest debt is to the late Professor John E. Arnold, who not only suggested that I develop a visual-thinking course at Stanford (a course that has been a major testing ground for this book) but also influenced me by his pioneering efforts to educate productive thinking. I also owe much to my students, whose difficulties and suggestions have been crucial to my education. My students taught me, for example, that their problems in sketching were often due to undeveloped powers of visual perception. This lesson convinced me that I should add concern with seeing to my original interest in idea-sketching.

Students also drew my attention to the fundamental relationship between idea-sketching and imagination. Some students claimed that they "had no imagination" and therefore no ideas to sketch, others revealed difficulty in departing from stereotyped ideas, and still others were frequently blocked in their generation of ideas. As I pondered how to deal educationally with these sorts of problems, a research colleague, psychologist Robert Mogar, pointed out the frequency of introspective accounts of "mind's-eye" imagery in the literature on creativity. Soon after, a number of other psychologists, referred to later in the section on *imagining*, taught me ways to invigorate and direct the inner sensory imagery that is the active model for idea-sketching.

Having added seeing and imagining to my initial concern with idea-sketching, I now searched for ways to integrate these three related visual activities, Professor Rudolf Arnheim, in an essay now expanded into an excellent book, suggested the term *visual thinking* to describe the interaction of seeing, imagining, and idea-sketching. Other authors, referenced later, gave me additional confidence that visual thinking had some basis in psychological knowledge and also suggested ideas that I could translate into experiential exercises. To those authors, and especially to those who have influenced me in unacknowledged conversations and letters, I extend my deepest appreciation. I also thank the following reviewers for reading the manuscript and giving many helpful suggestions: Vaughn P. Adams, Jack L. Alford, Jose Arguelles, William Bowman, Peter Z. Bulkeley, Jack Crist, Jay Doblin, W. Lambert

Gardiner, William Katavolos, Dean Myers, and Edward L. Walker.

Thanking people who influenced me by personal contact is more difficult. How does one acknowledge a crucial conversation or an off-hand comment that turned out to be a major lead? For showing me the value of experiential education, I especially thank Mike Murphy, Jack Downing, Claudio Naranjo, Laura Huxley, Alan Watts, Fritz Faiss, Ida Rolf, Bernie Gunther, George and Judy Brown, and the late Fritz Perls. For sharpening my interest in thinking strategies, I thank David Straus. For giving me insights into the process of idea-sketching as well as examples of idea-sketches, I thank graphic thinkers from many fields (credited with their sketches) and especially my fellow designers and design educators. For their support of my departures from the educational norm, I thank my friends and colleagues in Stanford's Design Division. For helping me to develop and test ideas that have found their way into this book, I thank my course co-teachers and teaching assistants. And not least, I thank my family for enduring my frustrated behavior while I was carrying out the ironic task of describing the following "experiences in visual thinking" with words.

I am grateful to the following sources for their permission to reproduce the figures indicated:

Figure 1–1: Photo courtesy of Phaidon Press Ltd.

Figures 2–2, 2–8, 12–2, 12–3, 13–4, 13–9, and 21–31: Photos courtesy of Bruce Thurston.

Figures 2–3 and 10–10: From *The Book of Modern Puzzles* by Gerald L. Kaufman, Dover Publications, Inc., New York, 1954. Reprinted through permission of the publisher.

Figures 2–4 and 2–5: Reprinted with the permission of J. D. Watson, Director, The Biological Laboratories, Harvard University.

Figure 2–6: Arthur Schatz, *Life Magazine,* © 1969 Time, Inc.

Figures 3–1, 3–2, and 3–3: Reprinted with the permission of The Industrial Relations Center, The University of Chicago.

Figures 3–4 and 3–8: From *Test Yourself* by William Bernard and Jules Leopold. Copyright © 1962 by William Bernard. Reprinted with the permission of the publisher, Chilton Book Company, Philadelphia.

Figures 3–5, 3–7, and 3–13: From Smith, I.M., *Spatial Ability.* San Diego, California: Robert R. Knapp, 1964. Reproduced with permission.

Figures 3–10 and 3–12: Reproduced by permission. Copyright 1947, © 1961, 1962 by The Psychological Corporation, New York, New York. All rights reserved.

Figure 3–11: From *Moray House Space Test 1,* by J. T. Bain and Colleagues, in consultation with W. G. Emmett and Professor Sir Godfrey H. Thomson (Education Department, Moray House, University of Edinburgh). Reprinted with permission of the Godfrey Thomson Unit for Academic Assessment and the University of London Press.

Figure 3–14: Reproduced by permission. Copyright 1947, © 1961 by The Psychological Corporation, New York, New York. All rights reserved.

Figure 3–15: Figure 36, Test 1 from Eysenck, H. J.: *Know Your Own I.Q.* Copyright © H. J. Eysenck, 1962. Reprinted by permission of the publisher, Penguin Books Ltd.

Figure 4–1: © 1971 by Jules Feiffer. Distributed by Publishers-Hall Syndicate.

Figure 5–1: Photo of Control Data's Digigraphic systems reprinted with the permission of the Control Data Corporation, Minneapolis, Minnesota.

Figure 8–2: *Postcard Study for Colossal Monument: Thames Ball,* 1967. Altered Postcard, 3½'' × 5½''. Collection: Carroll Janis, New York. Courtesy Sidney Janis Gallery, New York.

Figure 9–1: From Röttger and Klante, *Creative Drawing: Point and Line.* Copyright 1963 by Otto Maier Verlag, Germany. American Edition by Van Nostrand Reinhold Company, New York. Reprinted with the permission of the publisher.

Figure 9–3: Courtesy of Tseng Ta-Yu.

Figure 10–1: Cover photograph of the Proceedings of the Fourth Conference on Engineering Design, by Dr. Leonard Kitts, School of Art, Ohio State University, Columbus, Ohio. Reproduced with the permission of Professor R. C. Dean, Thayer School of Engineering, Dartmouth College, Hanover, New Hampshire.

Figure 10–3: From Wolf von Eckardt, *Eric Mendelsohn,* 1960, George Braziller, Inc. Reprinted with the permission of Mrs. Eric Mendelsohn.

Figure 10–6: Levi Strauss & Co./Honig-Cooper & Harrington.

Figure 10–7: From *Emilio Greco Sculpture and Drawings,* by J. P. Hodin. Copyright 1971 by Adams and Dart Publishers, Bath, Somerset, England. Reprinted by permission.

Figure 10–9: Figure 1–6 of *New Think* by Edward de Bono, © 1967, 1968 by Edward de Bono, Basic Books, Inc., Publishers, New York. Reprinted with permission of the publisher and Edward de Bono, author of *The Mechanism of Mind* (Simon & Schuster), *Lateral Thinking for Management* (American Management Association), and *Beyond Yes and No* (Simon & Schuster).

Figure 12–1: Reproduced from *Principles of Figure Drawing* by Alexander Dobkin. Copyright © 1960, 1948 by Alexander Dobkin.

200

Figure 12–4: Oliver Wendell Holmes. Detail from a drawing by David Levine in *Life Magazine*, October 1971. Reprinted with permission of David Levine.

Figure 13–1: Adapted from R. L. Gregory, *Eye and Brain*, McGraw-Hill Paperbacks. Copyright 1966 by R. L. Gregory. Reprinted with the permission of the publisher.

Figure 13–2: *False Perspective* by Hogarth, courtesy of Phaidon Press Ltd. *Relativity* by M. C. Escher, Collection Escher Foundation, Haags Gemeentemuseum, The Hague.

Figure 13–5: Photo courtesy of William Vandivert.

Figure 13–8: Drawing by Dürer courtesy of Phaidon Press Ltd.

Figure 14–1: From *How to Win Games and Influence Destiny* by Rick Strauss, Copyright © 1960 by Gryphon House, Box 1071, Pasadena, California 91102. Reprinted by permission.

Figure 18–1: Drawing by Cobean; © 1950, 1978 The New Yorker Magazine, Inc.

Figure 19–1: Reprinted from *Mazes* by Vladimir Koziakin. Copyright © 1971 by Vladimir Koziakin. Published by Grosset & Dunlap, Inc. Published in the United Kingdom and Commonwealth by Pan Books Limited, London. Reprinted by permission of the publishers.

Figure 20–1: Photo courtesy of United States Department of the Interior, National Park Service, Edison National Historic Site. Drawing of light bulb courtesy of General Electric Company, Cleveland, Ohio.

Figures 20–3, 21–49, and 21–50: Courtesy of Nancy Strube, Correlations, New York, and William Katavalos.

Figure 21–1: Photo courtesy of the French National Tourist Office, San Francisco, California.

Figure 21–2: Population map from the Department of Geography, University of Michigan. Photo courtesy of California Computer Products, Inc., Anaheim, California.

Figure 21–4: Drawn from C. K. Ogden and I. A. Richards, *The Meaning of Meaning*. Copyright 1923 by Harcourt Brace Jovanovich, Inc., New York. Published in the British Commonwealth by Routledge & Kegan Paul Ltd., London. Reprinted by permission of the publishers.

Figure 21–8: Pablo Picasso, *Horse and Bull*, early May 1937. Study for *Guernica*. Pencil on tan paper, 8⅞'' × 4¾''. On extended loan to The Museum of Modern Art, New York, from the artist's estate.

Figure 21–9: Redrawn, by permission of the publishers, from page 179 of *Language in Thought and Action*, Second Edition, by S. I. Hayakawa, copyright 1941, 1949, © 1963, 1964, by Harcourt Brace Jovanovich, Inc. Published in the British Commonwealth by George Allen and Unwin Ltd., London.

Figures 21–15 and 21–16: From Alexander, C., Ishikawa, S., and Silverstein, M., *A Pattern Language Which Generates Multi-Service Centers*. Copyright © 1968 by The Center for Environmental Structure. Reprinted by permission.

Figure 21–21: From Philip James, *Henry Moore on Sculpture*. Copyright by MacDonald & Co. Ltd., London. Reprinted by permission.

Figure 21–22: Copyright © 1961 by Henry Litolff's Verlag, care of C. F. Peters Corporations, 373 Park Avenue South, New York, NY 10016. Reprint permission granted by the publisher.

Figure 21–23: RSVP Diagram from *RSVP Cycles* by Lawrence Halprin, published by George Braziller, Inc. © 1969 Lawrence Halprin.

Figures 21–24, 21–27, and 21–44: Courtesy of Varian Associates, Palo Alto, California.

Figure 21–30: Reproduced by permission of Benjamin/Cummings Publishing Company, distributor of the Benjamin/Maruzen HGS Molecular Model Kits.

Figure 21–34: Courtesy of the Ford Archives, Henry Ford Museum, Dearborn, Michigan.

Figure 21–35: Conceptual drawing by architect Frederick Kiesler for the *Grotto for Meditation*. Reproduced by permission.

Figure 21–36: From *The Development by Paolo Soleri of the Design for the Cosanti Foundation, Arizona, U.S.A.* © Copyright 1964, Student Publications of the School of Design, North Carolina State of the University of North Carolina at Raleigh, Volume 14, Number 4, Sketch #112. Reprinted with the permission of the editors of the Student Publications of the School of Design and The M.I.T. Press, publisher of *The Sketchbooks of Paolo Soleri* by Paolo Soleri. Copyright 1971 by The M.I.T. Press.

Figure 21–42: Courtesy of Del Coates, Art School of the Society of A. & C.

Figure 21–43: Royal Collection, Windsor Castle, copyright reserved. Reproduced by gracious permission of Her Majesty Queen Elizabeth II.

Figure 21–46: From *The Development by Louis Kahn of the Design for the Second Capitol of Pakistan, Dacca*. © Copyright 1964, Student Publications of the School of Design, North Carolina State of the University of North Carolina at Raleigh, Volume 14, Number 3, Figure 16. Reprinted with the permission of the editors of the Student Publications of the School of Design.

Figures 21–53 and 21–54: Courtesy of Lockheed Aircraft Corporation.

Figure 21–55: From *The Development by Le Corbusier of the Design for l'Eglise de Firminy, A Church in France*. © Copyright 1964, Student Publications of the School of Design, North Carolina State of the University of North Carolina at Raleigh, Volume 14, Number 2, Model: December 1962. Reprinted with the permission of the editors of the Student Publications of the School of Design.

Finally, I thank the following sources for their permission to quote material in the text:

F. C. Bartlett, *Remembering*. Copyright by Cambridge University Press, New York. Reprinted with permission of the publisher.

J. S. Bruner, *On Knowing: Essays for the Left Hand*. Copyright 1962 by The Belknap Press of Harvard University Press, Cambridge, Mass. Reprinted with permission of the publisher.

Put Your Mother on the Ceiling, by Richard de Mille. Copyright © 1955, 1967 by Walker and Company. Reprinted with permission of the publisher, Walker and Company.

W. J. J. Gordon, *Synectics*. Copyright 1961 by Harper & Row, Publishers, Inc., New York. Reprinted with permission of the publisher.

Edward Hill, *The Language of Drawing*, © 1966. Reprinted by permission of Prentice-Hall, Inc., Englewood Cliffs, New Jersey.

A. Huxley, *The Art of Seeing*. Copyright 1942 by Harper & Row, Publishers, Inc., New York. Reprinted with permission of the publisher.

"Genius at Work" by Stanley Krippner and W. Hughes: Reprinted from *Psychology Today* Magazine, June, 1970. Copyright © 1970 by Ziff-Davis Publishing Company.

From *Imagination and Thinking* by Peter McKellar, Basic Books, Inc., Publishers, New York, 1957. Reprinted with permission of the publisher.

K. Nicolaides, *The Natural Way to Draw*. Copyright 1941, 1969 by Houghton Mifflin Company, Boston. Reprinted with permission of the publisher.

F. S. Perls, *Ego, Hunger, and Aggression*. Copyright 1966 by Random House, Inc., New York. Reprinted with permission of the publisher.

P. Picasso as quoted by Simonne Gauthier in *Look Magazine*, December 10, 1968. Reprinted by permission of Cowles Syndicate.

From *Metamorphosis* by Ernest Schachtel, Basic Books, Inc. Publishers, New York, 1959. Reprinted with permission of the publisher.

D. A. Straus in *Tools for Change: A Sourcebook in Problem-Solving*. Interaction Associates, Inc., 2637 Rose Street, Berkeley, California. November, 1971.

"For Phobia: A Hair of the Hound" by J. Wolpe: Reprinted from *Psychology Today*, June 1969. Copyright © 1969 by Ziff-Davis Publishing Company.

INDEX

203

207